*Routledge Revivals*

# Assessing the Demographic Impact of Development Projects

First published in 1992, *Assessing the Demographic Impact of Development Projects* based on studies in developing countries focuses on conceptual, methodological and policy issues related to development projects. It considers whether demographic effects can be assessed and why development planners should be interested in such an assessment. A.S. Oberai examines the extent to which economic and social changes generated by specific development interventions have influenced demographic behaviour in a particular context. He suggests how desired effects can be enhanced and undesired effects minimized by policy makers and planners in developing countries in order to deal with problems of population growth and its distribution. The major shortcomings of existing methodologies are identified, and the author indicates the future direction which research might take in order to be more scientifically valid and useful to policy makers. This book is a must read for scholars and researchers of development studies, political economy, and labour economy.

# Assessing the Demographic Impact of Development Projects

## Conceptual, methodological and policy issues

A. S. Oberai

First published in 1992
by Routledge

This edition first published in 2022 by Routledge
4 Park Square, Milton Park, Abingdon, Oxon, OX14 4RN

and by Routledge
605 Third Avenue, New York, NY 10017

*Routledge is an imprint of the Taylor & Francis Group, an informa business*



A Library of Congress record exists under ISBN: 041506841X

ISBN: 978-1-032-32174-5 (hbk)
ISBN: 978-1-003-31321-2 (ebk)
ISBN: 978-1-032-32176-9 (pbk)

Book DOI 10.4324/9781003313212]

# Assessing the demographic impact of development projects

## Conceptual, methodological and policy issues

A.S. Oberai

A study prepared for the International Labour Office within the framework of the World Employment Programme, with the financial support of the United Nations Population Fund (UNFPA)

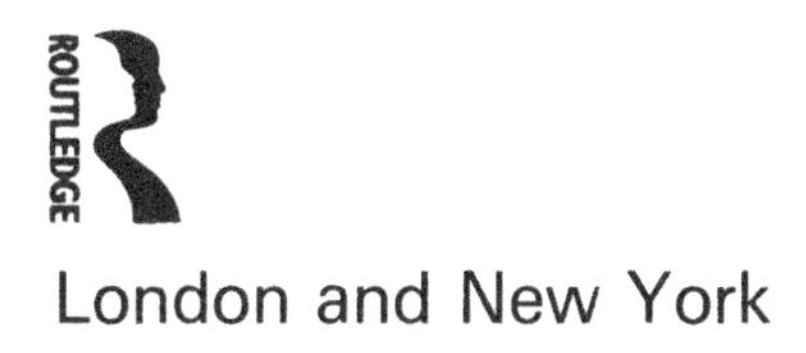


London and New York

First published 1992
by Routledge
11 New Fetter Lane, London EC4P 4EE

Simultaneously published in the USA and Canada
by Routledge
a division of Routledge, Chapman and Hall, Inc.
29 West 35th Street, New York, NY 10001



Typeset in Baskerville
by Pat and Anne Murphy, Highcliffe-on-Sea, Dorset
Printed in Great Britain by
Biddles Ltd, Guildford and King's Lynn



*British Library Cataloguing in Publication Data*
Oberai, A.S.
Assessing the demographic impact of development projects: conceptual, methodological and policy issues: a study prepared for the International Labour Office within the framework of the World Employment Programme.
1. Population. Effects of. Development projects
I. Title II. International Labour Office. World Employment Programme
304.6

ISBN 0-415-06841-X

*Library of Congress Cataloging-in-Publication Data*
Oberai, A.S.
Assessing the demographic impact of development projects: conceptual, methodological, and policy issues/A.S. Oberai.
p. cm.
'A study prepared for the International Labour Office within the framework of the World Employment Programme, with the financial support of the United Nations Population Fund (UNFPA)'.
Includes bibliographical references and index.
ISBN 0-415-06841-X
1. Developing countries - Population - Economic aspects - Case studies. 2. Economic development projects - Developing countries - Case studies. I. International Labour Office. II. World Employment Programme. III. United Nations Population Fund.
IV. Title.
HB884.O25 1991 91-10017
304.6'09172'4-dc20 CIP

# Contents

# Figures and tables

## FIGURES



## TABLES

**APPENDIX**

# Preface

This volume consists of a synthesis of country studies reviewing the demographic impact of development projects carried out in Bangladesh, Indonesia, the Philippines and Thailand. It also includes analysis of the demographic impact of development interventions in several other countries such as Brazil, Colombia, Costa Rica, India and Nigeria.

The study discusses conceptual, methodological and policy issues in assessing the demographic impact of development projects. In particular, it examines why development planners should be interested in assessing demographic effects and whether such effects can be assessed. It also examines whether and to what extent economic and social changes generated by specific development interventions such as credit and income-generating schemes for women, public works programmes, land settlements, rural electrification, irrigation schemes and integrated rural development programmes have influenced demographic behaviour (fertility, mortality and migration) in a particular context. At the same time it suggests how desired effects can be enhanced and undesired effects minimized by policy-makers and planners in developing countries in order to deal with problems of population growth and its distribution. An attempt is also made to identify the major shortcomings of the existing methodologies for assessing demographic impact and to indicate the future direction which research might take in order to be more scientifically valid and more useful to policy-makers. The study should therefore be of direct use to those engaged in population and development planning in Third World countries.

A number of persons have contributed to this volume. Country review studies on Bangladesh, Indonesia, the Philippines and Thailand were prepared by Dr A. Islam, Department of Statistics,

Dhaka University, and Dr M.A. Mabud, Planning Commission, Dhaka; Dr T.H. Hull and Dr V.J. Hull, International Population Dynamics Programme, Australian National University; Dr A.N. Herrin, School of Economics, University of the Philippines; and Dr S. Prasith-Rathsint, National Institute of Development Administration, Bangkok, respectively. I owe a deep sense of gratitude to the authors of the country review studies for their excellent work. Special thanks are also due to Eddy Lee, Hamid Tabatabai, Allan Parnell, W.R. Bohning and Richard Anker for their valuable comments on the earlier drafts of the chapters.

I wish to acknowledge the research assistance received from Dr D. Ghosh and Ms Suja Rishikesh. The work on typing the manuscript was handled by Mrs A. Eggleston, whose assistance is gratefully acknowledged.

I should also like to express my gratitude to the United Nations Population Fund (UNFPA) for their support of this work.

Needless to say, the views and opinions expressed in this volume reflect those of the author, and not necessarily those of the ILO or the UNFPA.

A.S. OBERAI
*Senior Economist,*
*Employment and Development Department*
*International Labour Office*

Chapter 1

# Introduction

It is now widely recognized among development planners that demographic factors such as migration, fertility and the acceptance of family planning are inter-related with socio-economic development. Few development planners believe that family planning programmes alone are sufficient to achieve national fertility objectives, particularly where the economic value of children is high because of limited income-earning opportunities and the need for old-age support for parents, and where infant and child mortality is high because of limited access to health and sanitation services. On the contrary, it is generally believed that until improvements in income and living conditions reduce the need to have large families, there will be little demand for contraception. There has therefore been a growing interest in many developing countries in implementing development programmes and educational and health strategies which will have a decisive impact on demographic behaviour, as well as contributing to economic growth and higher standards of living.

Many observers have also suggested that increases in the acceptance of family planning and reductions in fertility can be brought about more rapidly if population programmes are integrated into development projects at the community or area level. Because integrated population-development programmes at the community level enhance employment and income opportunities, particularly for women, and give people direct access to social and family welfare services, they are believed eventually to create an environment conducive to changing contraceptive and reproductive behaviour consistent with national population objectives. Several recent international conferences on population have also recommended that countries wishing to affect fertility give priority to linking family planning programmes with development activities.

Though most planners now agree that many of the development activities by themselves or in combination with population programmes will reduce fertility in developing countries, several important issues still remain. For example, what type of development activities will reduce fertility, by how much and in what time period will they do so, and what can governments do to accelerate the process of such change? Answers to such questions are extremely valuable for policy-makers in order to help shift the allocation of resources to development activities which have a more favourable demographic impact, as well as to enhance the effectiveness of integrated population-development programmes through appropriate changes in project design.

But it must also be recognized that a single-minded pursuit of a lone objective such as reduction in population growth may not always be desirable. Often there are serious conflicts between demographic and development objectives or between short-term and long-term welfare goals. In the case of land settlement programmes, for example, access to land may lead to greater demand for family labour and thus encourage higher fertility, particularly in the short term. Such short-term adverse consequences of settlement programmes on population growth need to be weighed against short-term or long-term benefits of increased agricultural production and reduced landlessness. However, if the only objective were to reduce the rate of growth of population then one might be tempted to prescribe increased mechanization in settlement areas in order to reduce the demand for children. But this may conflict with the policy of employment promotion. Thus in situations where there is a conflict between demographic and development objectives, such conflicts need to be identified and taken into account while judging the relative merits of alternative policies and programmes or when considering changes in project design so as to minimize the negative effects on population growth.

It is in this practical context that the ILO commissioned a number of studies reviewing the demographic impact of development projects in such countries as Bangladesh, Indonesia, the Philippines and Thailand. The purpose of the country case studies was to examine how development interventions such as credit and income-generating schemes for women, public works programmes, land settlement, rural electrification, irrigation schemes and integrated rural development programmes have influenced demographic behaviour through their effect on the socio-economic conditions of the communities concerned. Within this framework, a number of policy-related questions were examined:

1 Is a reduction in demographic pressure always desirable? If not, what are the trade-offs between population and development objectives?
2 What are the mechanisms through which development interventions affect demographic behaviour?
3 Which development interventions are more likely to have major direct and/or indirect demographic effects?
4 Under what circumstances and for which segments of the population are these effects likely to be greatest?
5 How best can the population objectives and/or programme components (e.g. population education and service delivery) be integrated into development projects?
6 How best can the demographic impact of development projects be measured and evaluated?

The present study reports the findings of the country case studies. In addition, it discusses the demographic impact of development interventions in several other countries such as Brazil, Colombia, Costa Rica, India and Nigeria.

The study is not intended to be a manual on how to design development projects so that they have, directly or indirectly, major demographic impacts in terms of reducing fertility and mortality, and slowing down rural to urban migration. Its major purpose is to discuss conceptual, methodological and policy issues in assessing the demographic impacts of development projects. As such, it examines why development planners should be interested in assessing demographic effects and whether such effects can be assessed. It also examines the major shortcomings of the existing methodologies for demographic impact assessment and indicates the future direction which research on demographic impact might take in order to be more scientifically valid and more useful to policy-makers in developing countries.

The study is divided into four chapters. After a brief introduction, the rest of this chapter is devoted to a discussion of demographic and development concerns in a historical perspective, the need for assessing the demographic impact of development projects and the major methodological problems encountered in the assessment of demographic impact. Chapter 2 reviews studies dealing with the demographic impact of selected development projects and critically evaluates the methodologies used by them. It also examines whether and to what extent economic and social changes generated by specific development projects have influenced demographic behaviour. Chapter 3 discusses the major shortcomings of existing methodologies for the assessment of demographic impact and suggests ways of improving them. It also cautions against evaluation attempts where data and other basic requirements for evaluation do not exist. Finally, Chapter 4 brings together the major findings of the study and highlights their policy implications.

Although the present study deals with the demographic effects of development projects in terms of fertility, mortality and migration, the focus of the study is on fertility and migration effects. This is not to suggest that mortality effects of development projects are relatively less important. In fact, reduction in infant and child mortality may be one of the first demographic effects of certain development projects and reciprocal effects between improved child survival and reductions in fertility are well known. The relative neglect of mortality effects in this study is, however, largely due to the fact that mortality has been less well studied in the existing studies of demographic impact assessment. This imbalance needs to be corrected in future research in this area.

## DEMOGRAPHIC AND DEVELOPMENT CONCERNS: HISTORICAL BACKGROUND

Shortly after the Second World War the literature on economic development in the Third World appeared to identify shortage of capital as the major stumbling-block in promoting economic growth. There was, however, some measure of optimism that the relatively abundant supply of labour could somehow, with appropriate economic policies, be mobilized to speed up capital formation. This approach was being put forward in the 1950s, even before population increase accelerated labour force growth. By and large, however, the hope that surplus labour could be effectively mobilized for capital formation failed to be realized.

When the United Nations proclaimed the 1960s as the First Development Decade, with economic growth as the engine of development, it was expected that employment creation, income distribution and other social objectives of development would be taken care of through a 'trickle-down' process. Hopes were also expressed in some quarters, based on the experience of developed countries, that with economic development demographic pressures in developing countries would eventually disappear. Though from a broad macro-economic point of view these hopes were perhaps justified, the experience of developed countries had shown that shifts from high to low levels of fertility and mortality occurred over periods ranging from 100 to 200 years and generally lagged behind shifts in technological development and per capita income. Therefore even if one were to take literally the argument that 'development is the best contraceptive', the Third World countries would never be able to develop sufficiently, within any reasonable period of time, to produce the automatic fertility reductions experienced in Europe.

By the end of the 1960s, the rapid decline in mortality in the early post-war period was already being translated directly into rapid population growth which was absorbing economic growth as fast as, or faster than, it could be achieved.[1] As such, there was little improvement in the economic well-being of the people. In spite of creditable output growth performance in many developing countries, the problems of poverty, unemployment and lack of basic social services were nowhere about to be solved. With disenchantment came reappraisal of the basic assumption of the development strategy that the rapid growth of output would eventually benefit the population via the 'trickle-down' process.

A growing consensus began to emerge in the international community that economic growth alone was not enough and that the demographic and social objectives of development should be addressed directly. The 'trickle-down' process of development was considered too slow and too uncertain. Attention in the 1970s therefore shifted to a development strategy that emphasized the need for direct provision of basic needs. Distinct sectoral objectives and plans in such areas of development as education, health and nutrition thus began to appear. The recognition of the relevance of population control to employment problems was also reflected in many development plans.

Several reasons were put forward as to why a reduction in population growth was likely to promote output and employment and thereby contribute to economic development. First, it was argued that lower fertility, and therefore reduced dependency, would lead to higher per capita savings and investment and an opportunity for faster growth in total income. Second, it was suggested that slower population growth would affect the composition of expenditure, especially investment and public expenditure. Slower growth in numbers would mean that a smaller proportion of this expenditure was required to provide each added person with the average amount of physical capital and social services, and so more would be available per person. Third, it was pointed out that a slower rate of growth of the labour force, together with increased investment, would reduce unemployment and improve wage levels and the distribution of income.

As the importance of the population factor in national development began to be more widely recognized, many of the developing countries began to initiate national family planning programmes (provision of contraceptive services together with information about contraception and child-spacing) as an integral part of development strategy. By the early 1970s, however, there was growing recognition that the narrow clinic-based family planning programmes alone were not sufficient to lower population growth to the desired level. The reason why these programmes did not prove as successful as was hoped for perhaps lies in the fact that they sought, by their very nature, to influence demographic factors, mainly fertility, through the supply side. Important though it was for family planning facilities to be made available to people, unless the people were willing to take advantage of them they could not prove to be effective.

On the demand side, there are several reasons why people in developing countries, particularly the poor, want large families. For example, most poor parents worry about who will take care of them when they are old or ill, One reason such parents look to children for help in disability and old age is the lack of safe alternatives.

As already mentioned, high infant and child mortality in many parts of the developing world is another reason for having many children. Parents usually feel the need to have many children to be sure that a few will survive. Where boys are considered to be more important than girls – say, for security in old age – parents often want to have several children to be sure that at least one son will survive.

Parents in developing countries are not always influenced by economic considerations alone, however. The social and political functions of large families are also important, especially in poor rural areas. A recent study has noted that for better-off farmers in Bangladesh, children represent opportunities for occupational diversification on the part of the family, and hence for expansion or consolidation of its local power; a large family is also an advantage in land disputes (World Bank, 1984).

Given the high demand for children in developing countries, many observers have suggested that a strategy for reducing population growth can only have a chance of success if family planning programmes are combined with policies and programmes which create a demand for family planning by reducing the demand for children.

## THE NEED TO ASSESS THE DEMOGRAPHIC IMPACT OF DEVELOPMENT PROJECTS

When allocating scarce resources, it is useful for planners to know the indirect demographic benefits (or costs) which may result from certain types of project and to consider how desired effects can be enhanced and undesired effects minimized. As Barlow (1982) has argued:

> irrigation projects, by raising rural incomes, may conceivably cause a rise in fertility. This could happen if, for example, children were viewed as a normal consumption good, or if the improved nutrition and health permitted by higher incomes lowered the frequency of miscarriages. A rise in fertility may eventually cancel out the economic improvement initially achieved. It seems logical to argue that project planners should be aware of the demographic impact of the development projects, and that the projects should be designed to optimise this impact (p. v).

Policy decisions based on cost-benefit analysis that does not take full account of interactions between demographic and socio-economic factors will therefore tend to be biased, constraining the achievement of development objectives. Industrialization programmes, for example, that do not consider the implications for the spatial distribution of economic activities and the consequent impact on the spatial distribution of population can indirectly lead to such adverse long-run consequences as regional income inequality, increased pressure on employment and basic social services in some urban centres and deteriorating income and employment prospects in the countryside.

While policy-makers can theoretically consider the potential impact of development on migration and fertility during the planning process, such explicit consideration rarely takes place, partly because of the absence of the empirical information required to quantify the demographic impact of development interventions (Barlow, 1982).

Very little is currently known about the demographic impact of most development projects and about the ways in which the country setting and the historical and cultural context intervene between the development projects and demographic outcomes. As a result, planners lack methodologies for projecting, monitoring and analysing the probable demographic consequences of development projects.

As a result, the projects that are undertaken often lead to demographic changes for which no provisions have been made. The opportunity is thus missed to incorporate into projects features which would increase desired demographic effects or reduce those considered negative. For planners, then, the usefulness of demographic impact assessment lies not only in investment allocation or priority ranking of development projects – which are often primarily based on socio-economic benefit and cost considerations – but also in the improvement of project design so that the impact on fertility will tend in the direction prescribed by national policy, downwards in almost all developing countries.

However, as already mentioned, in some situations there may be a genuine conflict between short-term and long-term policy objectives. In the short term, many income- and welfare-enhancing projects may actually lead to more demographic pressure, although in the long term, the demographic effect may be anti-natalist. In such cases, policy-makers and planners need to evaluate carefully the trade-offs between population and development objectives. Nevertheless, they may still be able to introduce changes in project design so as to minimize the short-term negative impact.

## PROBLEMS IN DEMOGRAPHIC IMPACT ASSESSMENT

Where attempts have been made to study the links between demographic and development processes, the macro-level approach of correlating a set of socio-economic indicators with population indices has been commonly adopted. But such an approach has serious disadvantages, especially for planning and policy purposes. First, it does not indicate the ways in which specific development interventions influence economic and demographic factors and so does not capture the complexities of the inter-relationships. Second, the impact of a specific development project often differs substantially from one geographical area or population subgroup to another, depending on the quality and quantity of development inputs, differences in the demographic composition of the target population and different behavioural response parameters. It is important therefore to focus analysis at the community, household and individual level where the impact of development policies and programmes is actually felt.

Although this micro-level analysis provides greater insights into the mechanisms through which development changes affect fertility and migration, it is often more difficult to generalize results on the basis of such an analysis. Moreover the methodological difficulties of conducting such an analysis are enormous. Perhaps it is largely due to these difficulties, rather than to a lack of interest in impact assessment, that researchers and policy-makers have so far paid little attention to assessing the demographic impact of development interventions or to examining how demographic concerns can best be incorporated into development activities at the community or household level.

In general, there are four major methodological problems in demographic impact assessment. First, development projects influence demographic behaviour only indirectly through changes in variables such as income, employment, health and education, and it is often difficult to trace out the demographic effects through changes in these intermediate variables. Second, since so many changes take place at the same time, it is difficult in practice to isolate changes due to a particular project or programme. Third, development projects are usually directed towards target populations with particular distinguishing characteristics. A study in India, for example, has shown that villages which receive electricity tend to be larger villages with secondary schools and agricultural services, and villages near cities. Thus to the extent that there is a selection bias in the choice of areas or households to receive a project, it is not correct to assume that observed demographic differences are due to the project rather than the result of other underlying differences. Finally, short-term effects are usually different from long-term effects. In the case of migration, for example, when an individual leaves the farming household the pressure resulting from the reduced supply of family labour may lead to an increase in the use of hired labour. But subsequently the household may adopt more labour-saving technology or sell part of the land (Oberai and Singh, 1983). In a demographic impact assessment study, measured impact will thus depend on the time-frame used and different periods of time between the initiation of projects and the evaluation of their effects will give different assessments.

Despite the methodological difficulties mentioned above, some researchers have attempted to assess the demographic impact of development interventions, largely using 'with and without' design, i.e. comparing areas receiving a project with those not receiving it. The results of these studies, along with a critical evaluation of the methodologies used by them, are discussed in the following chapter.

In reviewing impact assessment studies, the emphasis has been on assessment of the methodologies used in these studies rather than on their findings. The aim here has been to suggest, with the advantage of hindsight, some improvements in existing methodologies for future impact assessment.

Chapter 2

# Demographic impact of specific development projects: review of findings and methodologies used by impact assessment studies

The purpose of this chapter is to review micro-level impact assessment studies, examining whether and to what extent specific development interventions have influenced demographic behaviour in a given environment. In reviewing demographic impact assessment studies an attempt is first made to discuss the mechanisms through which a specific development intervention (electrification, irrigation, etc.) brings about changes in socio-economic and demographic factors. The empirical results presented in the studies, as well as the methodologies used by them, are then critically evaluated and their shortcomings highlighted. (Salient features of the studies reviewed in this section are also presented in summary form in Tables I to V in the Appendix.)

## RURAL ELECTRIFICATION

Rural electrification is expected to influence demographic behaviour indirectly through changes in such variables as income, employment, female labour force participation and general living conditions over time. Changes in these intermediate variables are usually felt at both the household and community level (Harbison and Robinson, 1985; Population Council, 1981).

At the household level, the availability and utilization of rural electrification, by itself and in combination with other development inputs, can affect female labour force participation in farm activities and the ability to use modern technology such as irrigation and mechanization in agriculture and home-based industry. These in turn can reduce the need to have more children to assist in household chores. The household work–leisure cycle can change, with household members spending more time in the evenings on activities such as reading and repair of productive assets. In other words, there can be more building up and maintaining of human and physical capital within the household. The change in daily work pattern, due to the changed technological (energy) base of the household, can lead to changes in the division of labour and roles within the family, particularly benefiting the female members of the household. Finally, the flow of information and outside influences through radio and television, and increased time spent on reading or studying, can give rise to changes in values and attitudes towards life and may thus reduce the desired family size.

At the community level, electrification (in conjunction with other development inputs such as irrigation, health, roads, etc.) can have a direct impact on employment and production patterns. It can stimulate rural industrialization and create new employment opportunities, especially for women. Electrification may also improve the efficiency of operation and utilization of many existing public service institutions such as schools and medical centres and encourage the establishment of new institutions where they do not exist. It can also encourage the setting-up of amenities such as cinemas and other recreational facilities, thereby increasing the attractiveness of the community. Finally, electrification can increase the flow of outside information and new ideas to the rural areas through radio and television and through the increased time available for viewing, listening and reading in the evening.

All these can be hypothesized to affect family planning and fertility. The new opportunities facing households tend to alter the perceived costs and benefits of children, leading towards a lesser demand for additional children. The actual fertility decline results from greater use of modern contraceptive methods available through an active family planning programme.

The effect of rural electrification on migration is usually ambiguous, however. On the one hand, increased employment opportunities and recreational facilities may help to increase the attractiveness of the rural areas, thereby reducing out-migration from the electrified rural areas or even giving rise to in-migration. On the other hand, increased educational facilities and flow of information through radio and television may in fact enhance the attractiveness of an urban life-style and give rise to increased rural–urban migration. Moreover, the decrease in the demand for family labour due to mechanization in agriculture may encourage some household members to migrate.

As mentioned earlier, the effect of rural electrification on demographic factors will depend on the mode, extent and length of utilization of electricity in the rural areas. If, for example, electricity is used at the community level only for street lighting, then it would have little effect on socio-economic and demographic factors. Similarly, if the majority of households in a rural area cannot afford to use electricity or choose not to use it for productive purposes, then the impact of rural electrification on demographic factors will be very small. The effect on fertility also takes a long time to be felt. In a recently electrified area, it will thus be unrealistic to look for evidence of a decrease in fertility as a result of electrification.

Piampiti *et al.* (1982) have assessed the demographic impact of a rural electrification project in north-eastern Thailand. The study used 'with and without' design and consisted of a sample survey of 5,000 rural households divided equally among electrified and non-electrified villages. The fieldwork was carried out from May to July 1981.

At the village level, separate regressions were run for electrified and non-electrified villages. Rural electrification was not included as an independent variable. The two equations simply showed what factors affected out-migration and crude birth rate in electrified and non-electrified villages. Thus no conclusion could be reached as to whether rural electrification had any direct effect on the dependent variables, namely out-migration and birth rate. It was nevertheless concluded from the analysis that whereas electrification was not significantly related to birth rate in the model, it had the effect of reducing out-migration.

In so far as path analysis is concerned (see Figure 2.1), the effect of rural electrification on recent fertility and migration was found to be indirect, passing through female labour force participation, husband's income, use of electric pump, desired family size, child mortality and contraceptive practice. With regard to recent fertility, the analysis indicated that the longer the household had used electricity, the higher was the husband's income, which was found to be positively related to the couple's contraceptive practice. Contraceptive practice, in turn, was negatively related to recent fertility. Analysis also showed that the longer the household had used electricity, the lower was the female labour force participation. Female labour force participation was, in turn, positively related to desired family size. In addition, desired family size was negatively related to the use of contraception.

As for migration, it was found that the use of electricity was

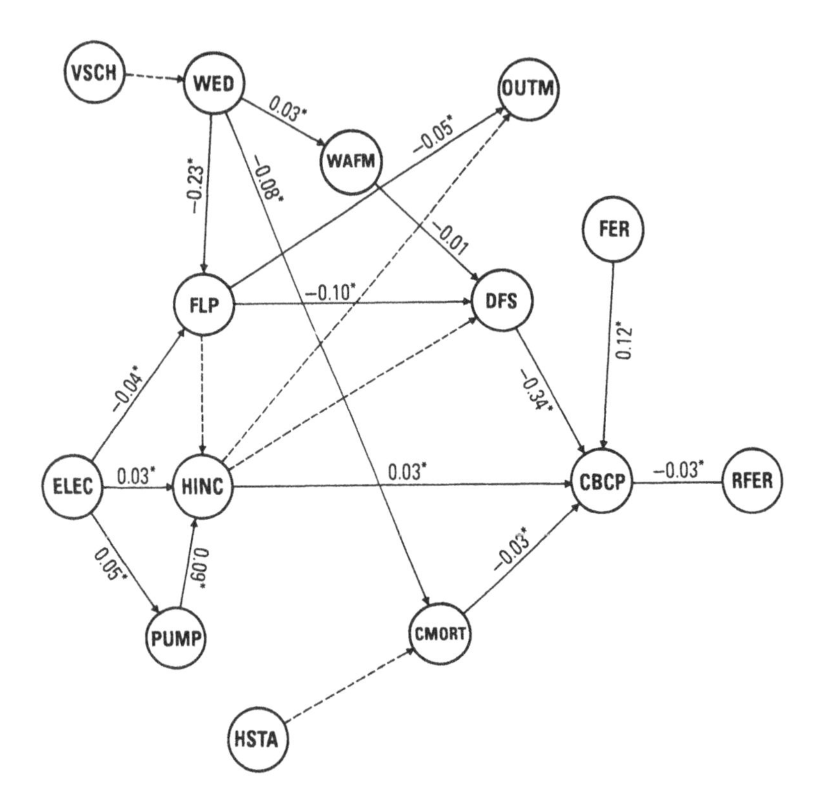


*Figure 2.1* Path diagram of the relationships between fertility, migration and other economic and demographic variables in an electrified rural village in Thailand

*Notes*:
* Significant at 1 per cent.
— — — Hypothesized but statistically not significant relationship.
________ Hypothesized and statistically significant.
VSCH = availability of village school; WED = wife's education; FLP = female labour force participation; ELEC = length of use of electricity; HINC = household income per capita; PUMP = use of electric water pump; HSTA = availability of health station; WAFM = wife's age at first marriage; CMORT = number of children aged under 5 years died in the household; DFS = desired family size; OUTM = number of household members out-migrated; CBCP = contraceptive birth control practice; FER = number of children ever born of the couple; RFER = number of children born during the last three years

*Source*: Piampiti *et al.* (1982), p. 124

negatively related to female labour force participation which in turn was negatively related to out-migration of the household members. However, it should be noted that the relationship between the use of electricity and female labour force participation is not in the expected direction. It was found to be negative instead of positive.

*Table 2.1* Availability and use of electricity and general characteristics of electrified and non-electrified villages in north-eastern Thailand, 1981

| *Particulars* | *Electrified villages (N = 60)* | *Non-electrified villages (N = 72)* |
|---|---|---|
| 1. Number of years since villages have had electricity (per cent) | | |
| 1–2 | 21.7 | – |
| 3–4 | 43.3 | – |
| 5–9 | 13.3 | – |
| 10–14 | 11.7 | – |
| 15 and more | 10.0 | – |
| | 100.0 | |
| 2. Percentage of households having: | | |
| (a) electricity | 70.9 | |
| (b) television | 11.7 | |
| 3. Percentage of households using electricity for: | | |
| (a) domestic consumption | 93.9 | |
| (b) productive use | 6.1 | |
| 4. Percentage of villages having: | | |
| (a) markets | 10.0 | 0 |
| (b) petrol stations | 25.0 | 9.7 |
| (c) health units | 40.0 | 13.9 |
| (d) family planning centres | 21.7 | 11.1 |
| (e) cinemas | 10.1 | 1.4 |
| (f) metalled roads | 65.0 | 29.2 |
| (g) primary schools | 68.0 | 65.0 |
| (h) secondary schools | 8.0 | 0 |

*Source*: Adapted from Piampiti *et al.* (1982), tables 3.4, 3.6, 3.7 and 3.16

From the point of view of the effect of electrification on fertility, a closer look at the data presented in Table 2.1 shows that the majority (65 per cent) of villages with electricity had had it for four years or less at the time of the survey. Since the time period required for electrification to affect fertility seems to be longer than four years, this could be one of the reasons why the researchers did not find any significant relationship between electrification and fertility.

Moreover, the data presented in Table 2.1 show that although a majority of households (70.9 per cent) had electricity at the time of the survey, only 6.1 per cent of them were using it for productive purposes (e.g. home industry, agricultural activities, animal raising). The study also observed that 'the average villager is unable to afford the luxury of the various amenities which electricity can provide' (Piampiti *et al.*, 1982, p. 67).

It seems therefore that the expected effect of electrification on fertility, which is largely felt through increased household income and production, increased use of recreational facilities, etc., has yet to work its way through in these villages.

With regard to migration, though, the result is as one would have expected. From the point of view of impact assessment it would have been more informative if a slightly different methodology had been adopted. As the data show, the electrified villages are much better served by business, health, recreational and educational facilities. In such a situation, 'before and after' methodology would have been more appropriate to indicate whether and to what extent the economic and social factors were affected by electrification. But such a methodology requires base-line information. In the absence of such information, the study should have adopted a 'then and now' strategy (based on official records on, say, number of industries, employment and turnover, number and types of school, number of health-related institutions, etc.) to complement the 'with and without' approach. This would have thrown more light on the effects of electrification on the intermediate factors.

Another study from Thailand (Pekanan, 1982) has produced rather interesting results. Adopting a 'with and without' methodology, the study selected married women aged between 15 and 44 years from 307 households in the electrified subdistrict of Hua Dorn and 302 from the non-electrified subdistrict of Chu Tuan in Ubon Ratchathani Province. The study concentrated on a comparison of recent fertility (i.e. over a four-year period, for which electricity was available for the first subdistrict) among women from electrified and non-electrified areas.

The results of multiple classification analysis and analysis of variance were taken to show that rural electrification was the most significant factor related to a decline in recent fertility. This conclusion is intriguing, since according to the study 'electricity has not brought about [significant] differences in the economic status of the people in Hua Dorn in the four years after access to electricity' (Pekanan, 1982, p. 27). Nor did the study find social conditions (defined in terms of leisure activities including use of radio and television, travel and expectation of children's education) any different in the two subdistricts, either before or after one of the villages was electrified. This is not surprising, since only 4.5 per cent of households in the electrified villages were found to be using electricity for economic purposes at the time of the survey.

*Table 2.2* Average number of live births and children ever born to married women in electrified and non-electrified villages in Ubon Ratchathani Province (Thailand), 1979

| | *Electrified villages* | | *Non-electrified villages* | |
|---|---|---|---|---|
| *Particulars* | *Users of contra-ceptives* | *Non-users of contra-ceptives* | *Users of contra-ceptives* | *Non-users of contra-ceptives* |
| 1. Average number of live births | | | | |
| (a) before electrification (1972–5) | 0.83 | 0.35 | 1.11 | 0.72 |
| (b) after electrification (1976–9) | 0.61 | 0.60 | 0.81 | 0.84 |
| 2. Rate of change in average number of live births during four years before and after electrification (%) | −26.5 | 71.4 | −27.0 | 16.7 |
| 3. Children ever born | 3.52 | 2.47 | 3.91 | 3.09 |

*Source*: Adapted from Pekanan (1982), tables 6 and 10

The data in Table 2.2 show that the average number of live births in the electrified villages was lower than that in the non-electrified villages during the four-year periods both *before* and *after* electrification. The rate of fall in the average number of live births over these periods (about 27 per cent) was also the same for women using contraceptives in the two groups of villages. However, the average number of live births over this period for non-users of contraceptives went up by 71.4 per cent in the electrified villages as compared to 16.7 per cent in the non-electrified villages.

On the whole, the data show that on average fertility has always been lower in the electrified villages than in the non-electrified ones.

One may thus suspect that the study has obtained a somewhat spurious statistical relationship between electrification and fertility, particularly since a time period of four years is not long enough for electrification to influence fertility behaviour.

In a study undertaken by Freedman *et al.* (1981) on a 1976 survey of Java and Bali in Indonesia, rural electrification was included as one community-level modernization variable (along with health, education and others). No positive relationship between the presence of electricity in a village and use of contraceptives was found, however. In fact contraceptive use was somewhat higher in villages without electricity (39.0 per cent) than in those with electricity (28.9 per cent).

The apparent anomaly of the Indonesian study is probably not as surprising as it first appears, and suggests a number of important questions for further study. The first likely explanation for the lack of relationship between electrification and fertility may lie in the very limited spread of electrification in Indonesia by the end of the 1970s and the large regional variations in contraceptive use and fertility. As McCawley pointed out in 1978, the distribution of electricity in the mid-1970s tended to be biased towards urban and suburban areas, and heavily favoured rich districts. Total coverage was low, with only 6 per cent of households nation-wide having electric lighting in 1971, and 14 per cent in 1980. Contraceptive use was biased towards the rural areas, on the other hand. Studies in the mid-1970s showed that in the early days of the family planning programme, services tended to be concentrated in rural areas (Hull, Hull and Singarimbum, 1977). Furthermore, there was evidence that well-to-do people were hesitant about using modern contraceptives, initially because of fear of side-effects, though they were using traditional methods of birth control such as herbs, massage and periodic abstinence. These factors may have confounded the results of Freedman *et al.*'s multivariate analysis (Hull and Hull, 1988).

Robinson *et al.* (1984) have evaluated the socio-economic impact of rural electrification in Bangladesh, for those areas electrified in 1980–1. The study involved a survey of six villages in each of four rural co-operative areas chosen randomly from a list of all villages in the areas. From the selected villages, 400 electrified and 200 non-electrified households were selected randomly. The methodology adopted was thus a 'with and without' one at the household level. The survey was carried out from January to April 1983.

The data presented in Table 2.3 shows that fertility, as measured by the number of children ever born, is higher in the electrified households (5.9) than in the non-electrified households (5.4). The data also show that about 82 per cent of the respondents in electrified households find family planning 'acceptable' as compared to about 73 per cent in the non-electrified households, although the percentage of respondents who actually practise family planning is almost identical in the two types of households. The desired family size is slightly lower among electrified (2.9) than among non-electrified (3.2) households.

*Table 2.3* Socio-economic and demographic characteristics related to electrified and non-electrified households in Bangladesh, 1983

| | *Household type* | |
|---|---|---|
| | *Electrified* | *Non-electrified* |
| 1. *Demographic characteristics* | | |
| (a) Children ever born | 5.9 | 5.4 |
| (b) Desired family size | 2.9 | 3.2 |
| (c) Percentage of respondents who: | | |
| (i) find family planning acceptable | 82.3 | 73.4 |
| (ii) practise family planning | 24.2 | 23.2 |
| 2. *Economic characteristics* | | |
| (a) Mean size of landholding (hectares) | 1.4 | 1.0 |
| (b) Households owning 2.4 hectares or more (%) | 30 | 11 |
| (c) Landless households (%) | 25 | 54 |
| 3. *Educational characteristics* | | |
| Education of husband (%) | | |
| (a) Illiterate | 16.0 | 41.9 |
| (b) Primary | 31.2 | 33.8 |
| (c) Secondary and higher | 52.8 | 24.2 |
| Education of wife (%) | | |
| (a) Illiterate | 48.6 | 77.0 |
| (b) Primary | 34.3 | 18.4 |
| (c) Secondary and higher | 17.2 | 4.6 |

*Source*: Adapted from Mabud and Islam (1987), pp. 85–8, tables 23 and 24

The slightly higher fertility observed among the electrified households can perhaps be explained by the relatively short time since the households received electricity, and also by continued problems in creating an effective delivery programme for family planning services in rural Bangladesh (Harbison and Robinson, 1985). Moreover, children ever born is not an appropriate measure for assessing the impact of electrification on fertility in the present context. An examination of fertility behaviour during the period *after* the experimental group of households were electrified would perhaps have been appropriate. Also, as Table 2.3 shows, the electrified households are, on average, richer and more educated than the non-electrified ones. This suggests that the comparison group selected was not appropriate for the investigation.

Robinson's study also used path analysis, which showed that rural household electrification increases family planning use by enhancing husband–wife communication and women's access to information, outside-home contacts, ability to earn income and desire for educational advancement for daughters (Mabud and Islam, 1987).

Herrin (1988) has assessed the impact of rural electrification on demographic factors in two areas (Misamis Oriental and Cagayan Valley) in the Philippines. Both the studies were conducted in 1980. The study in the province of Misamis Oriental included 18 villages from the western part of the province where electrification was introduced in 1971, and 18 villages from the eastern part where electrification was introduced in 1978. The latter group of villages were considered as a comparison group. The study therefore basically used 'with and without' design.

Four sets of surveys were conducted from September to November 1980. These included:

1 a household survey covering 1,402 households in 36 selected villages to determine the households' social, economic and demographic characteristics;
2 a community survey to determine the locational and socio-economic characteristics of the sample communities;
3 a user survey of business enterprises, educational and health units and other establishments to determine the type and extent of use of electrical equipment by such establishments;
4 an experience survey of village leaders, school officials, health personnel and other key persons to obtain from them personal observations and assessments of the social and economic changes associated with rural electrification.

Regression analysis was used to determine the effect of electrification and other factors on recent fertility, child mortality and family planning.

The definitions of the variables used in the estimation of fertility equations were as follows:

| Variable | Definition/measurement |
| --- | --- |
| *Dependent* | |
| FERT5 | Number of live births during the past five years. |
| FERT2 | Dummy variable (1 = if the woman has a live birth during the past two years; 0 = otherwise). |
| FPLANA | Dummy variable (1 = if the woman is currently using any family planning method; 0 = otherwise). |
| FPLANM | Dummy variable (1 = if the woman is currently using 'modern' family planning methods, that is, either the woman is using pills or IUD, or either she or her husband is sterilized; 0 = otherwise). |
| *Independent* | |
| *Personal characteristics* | |
| AGEW | Age of wife in years as of last birthday. |
| AGEWSQ | The square of AGEW to allow for non-linearities. |
| AGEM | Age at marriage in completed years. |
| EDW | Educational attainment of wife, measured as the highest grade of schooling completed. |
| PPARITY5(2) | Number of children ever born five (two) years prior to the date of interview. |
| PPARITY5(2)SQ | The square of PPARITY5(2) |
| PWAGEW (PWAGEH) | Natural logarithm of the hourly wage rates of wife (husband) predicted from husband's (wife's) background characteristics, locational characteristics and rural electrification. |
| *Household characteristics* | |
| HOUSE | Dummy variable (1 = if house is made of light construction materials; 0 = if made of medium or heavy materials). |
| OWNHOUSE | Dummy variable (1 = if the household owns house; 0 = otherwise). |
| OWNLOT | Dummy variable (1 = if the household owns home-lot; 0 = otherwise). |
| OWNLAND | Dummy variable (1 = if household owns agricultural land; 0 = otherwise). |
| YCRES | Dummy variable (1 = if the couple has resided continuously in the area since 1971; 0 = otherwise). |
| *Locational/development variables* | |
| DISTPOB | Distance of rural community from municipal town centre in kilometres. |
| DISTTHWY | Distance of community from national highway in kilometres. |
| DISTCGY | Distance of community from Cagayan de Oro, the provincial capital in kilometres. |
| YAELEC | Number of years since the community was electrified. |

*Table 2.4* Estimated regression equations for the determinants of live births during the past five and two years among married women in Misamis Oriental (the Philippines), 1980

| Independent variables | *Live births during the past* — *Five years*: *Mean (standard deviation)* | *Five years*: *Coefficient (t-values)* | *Two years*: *Mean (standard deviation)* | *Two years*: *Coefficient* |
|---|---|---|---|---|
| AGEW | 35.85 | 0.027 | 34.66 | 0.171[a] |
| | (7.22) | (0.56) | (7.76) | (6.83) |
| AGEWSQ | — | −0.002[a] | — | −0.003[a] |
| | | (−2.62) | | (−7.75) |
| PPARITY5 | 4.20 | −0.052 | — | — |
| | (2.95) | (−1.24) | | |
| PPARITY5SQ | — | 0.009[a] | — | — |
| | | (2.92) | | |
| PPARITY2 | — | — | 4.42 | −0.358[a] |
| | | | (3.09) | (−19.33) |
| PPARITY2SQ | — | — | — | 0.028[a] |
| | | | (19.80) | |
| AGEM | 19.90 | 0.032[a] | 20.14 | −0.015[b] |
| | (3.91) | (3.02) | (4.00) | (−2.37) |
| EDW (5–7) | 0.47 | −0.129 | 0.45 | −0.050 |
| | (0.50) | (−1.37) | (0.50) | (−1.09) |
| EDW (8–11) | 0.24 | −0.069 | 0.25 | −0.071 |
| | (0.42) | (−0.42) | (0.44) | (−1.29) |
| EDW (12+) | 0.09 | −0.134 | 0.10 | −0.026 |
| | (0.29) | (−0.52) | (0.30) | (−1.28) |
| HOUSE | 0.67 | 0.257[a] | 0.69 | 0.070 |
| | (0.47) | (3.83) | (0.46) | (1.54) |
| OWNHOUSE | 0.87 | −0.240[b] | 0.86 | −0.084 |
| | (0.33) | (−2.60) | (0.35) | (−1.44) |
| OWNLOT | 0.35 | 0.010 | 0.34 | −0.024 |
| | (0.48) | (0.13) | (0.47) | (−0.49) |
| OWNLAND | 0.26 | −0.065 | 0.25 | −0.088[c] |
| | (0.44) | (−0.83) | (0.43) | (−1.67) |
| YCRES | 0.86 | −0.208[b] | 0.86 | −0.107[c] |
| | (0.35) | (−2.44) | (0.35) | (−1.92) |
| COASTAL | 0.33 | −0.065 | 0.33 | −0.063 |
| | (0.47) | (−0.69) | (0.47) | (−0.88) |
| INLAND | 0.33 | 0.194 | 0.33 | 0.095 |
| | (0.47) | (1.53) | (0.47) | (1.01) |
| DISTPOB | 2.00 | −0.042 | 2.04 | 0.004 |
| | (2.82) | (−0.96) | (2.93) | (0.14) |
| DISTHWY | 1.29 | 0.009 | 1.33 | −0.021 |
| | (2.87) | (0.19) | (3.00) | (−0.67) |
| DISTCGY | 48.13 | 0.002 | 48.14 | 0.001 |
| | (16.30) | (0.47) | (16.33) | (0.30) |
| PWAGEW | −0.02 | −0.130 | 0.02 | 0.061 |
| | (0.57) | (−0.82) | (0.59) | (0.41) |
| PWAGEH | 0.66 | 0.155 | 0.66 | 0.077 |
| | (0.40) | (1.18) | (0.40) | (0.83) |

*Table 2.4* – continued

| *Independent variables* | *Live births during the past* | | | |
|---|---|---|---|---|
| | *Five years* | | *Two years* | |
| | *Mean (standard deviation)* | *Coefficient (t-values)* | *Mean (standard deviation)* | *Coefficient* |
| YAELEC | 4.30 (3.7) | −0.028[b] (−2.19) | 4.32 (3.70) | −0.025[a] (−2.81) |
| Constant | 2.088 | | −0.477 | |
| R−2 | | 0.304 | | 0.374 |
| N | 1077 | | 1207 | |
| Mean of the dependent variable | | 1.210 | | 0.587 |
| (Standard deviation) | | 1.120 | | 0.822 |

*Notes*: [a] Significant at 1 per cent
[b] Significant at 5 per cent
[c] Significant at 10 per cent

*Source*: Herrin (1988), Section III, tables 2 and 3

The results of the regression analysis, presented in Table 2.4, show that the net effect of rural electrification on fertility over the past five years, as indicated by the coefficient of the variable relating to years since using electricity (YAELEC), is negative and significant. As an intervening mechanism, the exogenous change in wage rates measured by the predicted wife's and husband's wage rates (PWAGEW and PWAGEH), is not statistically significant at the 10 per cent level. The data, although not presented here, also show that rural electrification significantly increases the wage rates of both husbands and wives, but this does not appear to influence current fertility significantly. However, when the variable relating to the education of the wife (EDW) is excluded from the regressions due to its high correlation with the variable relating to the predicted wife's wage rate (PWAGEW), the coefficients for the wage rates gain statistical significance and in the hypothesized direction (Herrin, 1988, Section III, Tables 2 and 3).

With regard to live births during the two years prior to the survey, the regression coefficient for the variable relating to years since using electricity (YAELEC) is again negative and significant. Similar observations can be made with respect to the effect of the wage variables as in the case of the regression on live births during the past five years.

*Table 2.5* Estimated regression equations for the determinants of current use of family planning methods in Misamis Oriental (the Philippines), 1980 (t-values are given in brackets)

| *Independent variables* | *Mean (standard deviation)* | *Current use of family planning* | |
|---|---|---|---|
| | | *Any method Coefficient* | *Modern method Coefficient* |
| AGEW | 34.18<br>(8.02) | 0.065[a]<br>(3.94) | 0.035[b]<br>(2.41) |
| AGEWSQ | – | −0.001[a]<br>(−4.19) | −0.001[a]<br>(−2.83) |
| PPARITY | 4.85<br>(2.96) | 0.059[a]<br>(3.21) | 0.046[a]<br>(2.82) |
| PPARITYSQ | – | −0.005[a]<br>(−3.57) | −0.004[a]<br>(−2.94) |
| AGEM | 20.31<br>(5.08) | −0.004<br>(−1.33) | −0.004<br>(−1.39) |
| EDW (5–7) | 0.45<br>(0.50) | 0.114[a]<br>(3.06) | 0.104[a]<br>(3.17) |
| EDW (8–11) | 0.26<br>(0.44) | 0.184[a]<br>(4.27) | 0.061<br>(1.62) |
| EDW (12+) | 0.11<br>(0.31) | 0.292[a]<br>(5.04) | 0.124[b]<br>(2.42) |
| HOUSE | 0.69<br>(0.46) | −0.050<br>(−1.57) | −0.041<br>(−1.47) |
| OWNHOUSE | 0.84<br>(0.37) | −0.023<br>(−0.58) | 0.037<br>(1.07) |
| OWNLOT | 0.33<br>(0.47) | 0.013<br>(0.38) | 0.063[b]<br>(2.10) |
| OWNLAND | 0.25<br>(0.43) | −0.023<br>(−0.64) | −0.026<br>(−0.82) |
| YCRES | 0.86<br>(0.35) | −0.044<br>(−1.12) | −0.057<br>(−1.63) |
| COASTAL | 0.33<br>(0.47) | −0.078<br>(−1.54) | −0.045<br>(−0.100) |
| INLAND | 0.33<br>(0.47) | 0.035<br>(−0.86) | 0.072[b]<br>(−1.97) |
| DISTPOB | 2.06<br>(2.96) | 0.058[b]<br>(2.83) | 0.035[b]<br>(1.96) |
| DISTHWY | 1.33<br>(3.03) | −0.070[a]<br>(−3.54) | −0.043[b]<br>(−2.47) |
| DISTCGY | 48.11<br>(16.31) | −0.004[a]<br>(−2.76) | 0.001<br>(0.60) |
| YAELEC | 4.30<br>(3.70) | 0.010[c]<br>(1.76) | 0.017[a]<br>(3.26) |
| Constant | | −0.539 | −0.405 |
| $R^{-2}$ | | 0.113 | 0.071 |
| N | | 1257 | 1257 |
| Mean of the dependent variable<br>(Standard deviation) | | 0.432<br>(0.495) | 0.240<br>(0.495) |

*Notes*: [a] Significant at 1 per cent
[b] Significant at 5 per cent
[c] Significant at 10 per cent

*Source*: Herrin (1988), Section III, table 4

As far as the effect of rural electrification on family planning is concerned, the regression results presented in Table 2.5 show that the variable relating to years since using electricity (YAELEC) is positively and significantly related to both the variables relating to family planning using any method (FPLANA) and family planning using 'modern' methods (FPLANM). These results are consistent with those of the fertility equations.

In addition to the possible effect of rural electrification on fertility and family planning through its influence on wage rates, Herrin's study also suggested that aspirations for new lifestyles which are not consistent with continued high fertility may be an important factor in changing fertility and family planning behaviour.

In contrast to the above results, the study conducted in Cagayan Valley failed to show any relationship between electrification and demographic variables. The survey conducted in 1980 (five years after the provinces were electrified) included 28 villages. Of these 12 were electrified and 16 non-electrified. In an effort to eliminate selection bias, the non-electrified villages were chosen using the same criteria as used for original selection of villages for electrification. The survey design used was similar to that in the Misamis Oriental study. The household survey covered 600 households in both electrified and non-electrified villages.

The definitions of the variables used in the estimation of fertility equations were as follows:

| *Variable* | *Definition/measurement* |
|---|---|
| *Dependent* | |
| CFERT 76 | Dummy variable indicating whether or not the women had a live birth after 31 December 1976, i.e. a year after the area was electrified. |
| *Independent* | |
| AELEC | Area electrification status (1 = if community is electrified; 0 = otherwise). |
| AGEW | Age of wife in years. |
| BGY | Type of area (1 = if sample area is a rural village; 0 = otherwise). |
| CGY | Cagayan (1 = if sample area is located in Cagayan Electric Co-operative (CAGELCO I); 0 = otherwise). |
| CS–76 | Total number of living children as of 13 December 1976. |
| DISTPOB | Distance of household from municipal *población* in kilometres. |
| DISTHWY | Distance of household from national highway in kilometres. |
| DISTPC | Distance of household from provincial capital in kilometres. |

*(continued overleaf)*

| *Variable* | *Definition/measurement* |
|---|---|
| EDW | Educational attainment of wife, measured as the highest grade of schooling completed. |
| HOUSE | Type of housing materials (1 = if house is made of light materials; 0 = otherwise). |
| OCCH | Occupation of husband (1 = if husband is a farmer; 0 = otherwise). |
| OWNLAND | Ownership of agricultural land (1 = if household own agricultural land; 0 = otherwise). |
| EWAGEW | Estimated natural logarithm of female hourly wage rate obtained from paid employment. |
| EFH | Estimated natural logarithm of total household income from all sources less the money contribution of the wife from paid employment (past 12 months). |
| EHELEC | Estimated household electrification status (1 = if the household is electrified; 0 = otherwise). |

Regression results are presented in Table 2.6. As shown in the table, both household and area electrification were included among independent variables but neither was found to be statistically significant.

These apparently contradictory results as to the effect of rural electrification on demographic factors in the two regions in the Philippines can perhaps be explained in terms of the prevailing socio-economic conditions in these regions. In the Misamis Oriental region rural electrification was part of a large-scale agro-industrial development and many households were given financial help towards meeting the cost of installing electricity by co-operatives and local and national government. In Cagayan Valley, on the other hand, the limited supply of electricity available did not encourage development of any large-scale enterprise. Also, given the relatively high cost of electricity in this region, the use of electricity was not widespread. For example, only 27 per cent of households and only one health unit in the electrified area had electricity.

Another study carried out in 1980 by ter-Wengel (1985) examined the effect of electrification on migration decisions of persons aged 15–29 using data for the period 1970–9 from Villa Gómez and El Peñón, two villages in Northern Cundinarmaca, Colombia. These two villages were chosen because of their similarities in socio-economic and demographic terms. The only difference was that while El Peñón was fully electrified in 1976, Villa Gómez, which was only partially electrified, extended its school programme by two years in 1977–8. The partial electrification in Villa Gómez meant 'having electricity from the municipal plant when it worked, which was often not the case, and then only from 6.00 p.m. to 9.00 p.m.' (ter-Wengel, 1985, p. 56).

*Table 2.6* Estimated regression equation for the determinants of current fertility (since 1976) among married women in Cagayan Valley (the Philippines), 1980 (t-values are given in brackets)

| *Independent variable* | *Mean (standard deviation)* | *Current fertility (CFERT 76)* |
|---|---|---|
| AGEW | 34.476 | −0.036[a] |
| | (8.233) | (8.287) |
| EDW | 7.077 | −0.030[b] |
| | (3.740) | (1.994) |
| CS–76 | 3.557 | 0.010 |
| | (2.629) | (0.931) |
| EWAGEW | −5.988 | 0.124[b] |
| | (8.191) | (2.142) |
| EFH | 6.585 | 0.101 |
| | (4.409) | (1.059) |
| OCCH | 0.610 | 0.174[c] |
| | (0.488) | (1.720) |
| OCCH × OWNLAND | 0.108 | −0.087 |
| | (0.311) | (1.132) |
| HOUSE | 0.632 | 0.021 |
| | (0.483) | (0.192) |
| BGY | 0.670 | −0.174[c] |
| | (0.471) | (1.844) |
| BGY × DISTPOB | 2.020 | 0.002 |
| | (2.185) | (0.226) |
| DISTHWY | 0.612 | 0.092[b] |
| | (1.564) | (2.143) |
| DISTPC | 44.405 | 0.002[b] |
| | (30.033) | (2.119) |
| CGY | 0.584 | 0.085 |
| | (0.493) | (1.153) |
| AELEC | 0.211 | −0.068 |
| | (0.409) | (0.927) |
| EHELEC | 0.348 | −0.347 |
| | (0.201) | (0.709) |
| Constant | | 1.413 |
| $R^{-2}$ | | 0.277 |
| N | | 454 |
| Mean of the dependent variable | 0.555 | |
| (Standard deviation) | | (0.497) |

*Notes*: [a] Significant at 1 per cent
[b] Significant at 5 per cent
[c] Significant at 10 per cent

*Source*: Herrin (1988), Section IV, table 11

Household interviews were carried out in the two villages, with special attention paid to questions on migration. Logit analysis was used to estimate the following regression equation for different years:

$$M_i = b_o + b_1 \text{ vill} + b_2 (\text{vill} \times \text{Age } 15\text{–}19)_i + b_3 \text{ RELSHIP}_i$$
$$\text{for } i = 1970, \ldots\ldots, 1979$$

where $M_i$ = migration, a dummy variable (0 = if the person remains in the village; 1 = if he out-migrated in year $i$);
Vill = dummy variable indicating the village;
1 = Villa Gómez, 2 = El Peñón;
(Vill × Age 15–19) = interaction term designed to measure the effect of the extension of the school programme in Villa Gómez, a dummy variable (1 = if a person aged 15–19 resides in El Peñón; 0 = otherwise);
$\text{RELSHIP}_i$ = a dummy variable (1 = if a previous member of the household had emigrated to the destination area of migrant in the i'th year; 0 = otherwise).

The estimated logit coefficients were then used to determine the probabilities of out-migration from the two villages for the years 1970–9. Table 2.7 shows that the probabilities of out-migration for the 20–29 age group increase in Villa Gómez (the non-electrified village) over time relative to those of El Peñón, particularly after the electrification of the latter in 1976.

These results suggest that electrification may indeed have played an important role in the retention of population. With respect to education, the data, although not presented here, convey the impression, though less persuasively, that the extension of the school programme may have helped in retaining young people 15–19 years old.

This 'with and without' design underlines the problem of finding a suitable 'comparison' group. It is not clear whether for the purpose of this study the village with the extended education programme or the village with electrification was treated as the control. On the basis of the statistical estimation reported, it seems that the expansion of education has had less effect in controlling migration than electrification. It is to be noted that the expansion of education in Villa Gómez 'may have influenced the retention of the population aged 15 to 19' (ter-Wengel, 1985, p. 58). However, since we do not know, and the choice of time period (only two years after the expansion of education was introduced) does not permit us to observe whether people of this age group, after receiving more education, migrated out or not, it seems that the project was carried out a little too soon to study the effect of education and electrification on out-migration.

*Table 2.7* Probabilities of out-migration of persons aged 15–19 and 20–29 from Villa Gómez and El Peñón (Colombia), 1970–9

| | *Persons aged 15–19* | | | | *Persons aged 20–29* | | | |
|---|---|---|---|---|---|---|---|---|
| | *Individuals without previously out-migrated relatives* | | *Individuals with previously out-migrated relatives* | | *Individuals without previously out-migrated relatives* | | *Individuals with previously out-migrated relatives* | |
| *Year* | *Villa Gómez* | *El Peñón* | *Villa Gómez* | *El Peñón* | *Villa Gómez* | *El Peñón* | *Villa Gómez* | *El Peñón* |
| 1970 | 0.9 | 6.1 | 1.8 | 11.5 | 0.9 | 14.2 | 1.8 | 24.8 |
| 1971 | – | – | – | – | – | – | – | – |
| 1972 | 2.6 | 4.0 | 4.7 | 7.2 | 2.6 | 10.5 | 4.7 | 17.8 |
| 1973 | 2.7 | 1.8 | 3.8 | 1.6 | 2.7 | 5.7 | 3.8 | 8.0 |
| 1974 | 2.0 | 1.4 | 9.6 | 6.8 | 2.0 | 7.4 | 9.6 | 29.2 |
| 1975 | 2.6 | 0.7 | 19.8 | 5.9 | 2.6 | 1.4 | 19.8 | 11.8 |
| 1976 | 6.1 | 2.9 | 19.9 | 10.2 | 6.1 | 3.5 | 19.9 | 12.1 |
| 1977 | 2.3 | 1.6 | 17.9 | 12.9 | 2.3 | 3.3 | 17.9 | 23.4 |
| 1978 | 9.8 | 2.6 | 19.5 | 5.3 | 9.8 | 2.6 | 19.5 | 5.6 |
| 1979 | 6.8 | 6.9 | 20.0 | 21.0 | 6.8 | 2.2 | 20.8 | 7.6 |

*Source*: ter-Wengel (1985), table 7

## IRRIGATION

Like rural electrification, irrigation affects demographic factors indirectly through a number of intermediate variables such as land utilization and land use pattern, income and productivity and labour force participation, at both the household and the community level.

At the household level, an irrigation system can increase the productivity of land by making multiple cropping possible, or by making hitherto barren land fit for cultivation. As development of an irrigation system is usually related to the availability of other development inputs, such as electricity, roads, health services and schools, the utilization of these services in conjunction with the irrigation system can enhance the increased demand for family labour and female labour force participation. Increased income and wealth, as a result of the increased agricultural productivity that irrigation can bring about, provide households with the ability to use electricity, education, health and recreational facilities. All these can affect ideal family size, fertility and migration.

At the community level, the availability of an irrigation system, given a certain size and nature of utilization, can affect the pattern of land use through the possible introduction of multiple cropping and also by helping to bring new land under cultivation – thereby increasing the ratio between cultivable land and population. An irrigation system can also reduce the element of uncertainty related to agricultural production in areas which otherwise have to depend on rainwater. In addition, the construction of irrigation projects is often tied to electrification, which, combined with higher income from increased agricultural production, may generate new agricultural technologies and transfer of resources to non-agricultural activities (Kocher, 1973).

An increase in income and reduced uncertainty about it can affect both migration and fertility. The increased well-being of the community can also help to generate new recreational and leisure activities, thus making the community more attractive; this should, in turn, reduce out-migration.

As far as fertility is concerned, the effects of irrigation in giving rise to a reduction in the demand for children will lead more pronouncedly to an actual reduction in fertility when people have access to family planning services.

But whether and to what extent irrigation will affect demographic variables, both at the household and at the community level depends on its duration of operation, its rate of utilization and the way it affects the economic and social conditions in the areas it serves. If an irrigation system benefits only 'rich' households and leads to concentration of land ownership, then the effect of the irrigation system in reducing fertility and out-migration will be less than where benefits are more equitably distributed.

An assessment of the demographic impact of the Girna Irrigation Project in Maharashtra State of India was carried out by Mukerji *et al.* (1986). For the study 100 villages, 50 experimental (i.e. villages which have benefited from the project) and 50 control (i.e. villages which have not benefited) were selected. The study was based on existing data, the main sources of demographic data being the two censuses of 1971 and 1981. The analysis was carried out at the level of the *taluka* (subdistrict), since village-level data were not available.

On the basis of indirectly estimated crude birth rates, the study found fertility to have continued at a high level throughout the ten-year period both in the experimental and in the control areas. In particular, the *taluka* which benefited the most from the irrigation project had the highest birth rate. On the other hand, the trend in annual birth rates as computed from birth registration data presented in Table 2.8 was markedly downwards. In the experimental villages they fell by 12.7 and 9.4 points and in the control villages by 7.5 and 9.5 points during the period 1971–81.

As far as migration is concerned, on the basis of population density figures over the ten-year period the study concluded that the experimental villages had gained population through in-migration.

*Table 2.8* Estimated annual crude birth rates in irrigated and non-irrigated *talukas* in Maharashtra (India), 1971 and 1981

| *Taluka (subdistrict)* | *1971* | *1981* |
|---|---|---|
| *Irrigated* | | |
| Erandol | 32.9 | 20.2 |
| Bhadgaon | 33.0 | 23.6 |
| *Non-irrigated* | | |
| Amalner | 35.5 | 28.0 |
| Parola | 35.4 | 25.9 |

*Source*: Adapted from Mukerji *et al.* (1986), table 2

The main drawback of this study is that it had to depend on data not specifically collected with the particular needs of the research project in mind. To assess the effects of fertility and migration the authors had to depend on indirect estimates which, in the case of fertility, yielded somewhat contradictory results. With regard to migration, the authors themselves pointed out that the poor quality of Civil Registration data did not permit them to carry out any in-depth analysis.

Another study which has examined the effect of an irrigation project on demographic factors relates to the Lam Pra Plerng irrigation dam in north-eastern Thailand. The study was carried out by Prasaratkul *et al.* (1985).

Three villages in the area irrigated by the dam were selected as 'experimental' areas and two villages outside the irrigated zone were selected as control areas. Experimental and control areas were selected adjacent to one another because it was intended 'to keep the effect of other development inputs the same as far as possible' (Prasaratkul *et al.*, 1985). The study used a combination of sample survey and anthropological approaches for the collection of information on population change and development in the sampled areas.

With regard to migration, the data presented in Table 2.9 show that two out of the three irrigated villages lost population through migration. The authors of the study argue that in these two villages the presence of irrigation helped to increase agricultural productivity, which in turn caused land prices to increase, thereby cutting off 'possibilities for poor farmers from adjacent rural areas to migrate in'. Though the irrigation project might have increased demand for agricultural labour, this demand seems to have been met by the considerable underemployment that existed in the area before. The only irrigated village which experienced net in-migration was not, in fact, so much affected by irrigation since, unlike the other two villages, where 100 per cent of the families were engaged in farming, only 50 per cent of the families in this village were farmers.

*Table 2.9* In- and out-migration in irrigated and non-irrigated villages in the Lam Pra Plerng project area (Thailand), 1975–9 (no. of persons)

| *Villages* | *In-migration* | *Out-migration* | *Net migration* | *Net migration as percentage of 1975 population* |
|---|---|---|---|---|
| *Irrigated* | | | | |
| Suam Hohm | 6 | 47 | −41 | −10.2 |
| Nok Ok | 71[a] | 44 | 27[a] | 3.5[a] |
| Ban Kok | 24 | 75 | −51 | −7.7 |
| *Non-irrigated* | | | | |
| Khun Lakorn | 6 | 44 | −38 | −11.3 |
| Nong Yai Ray | 10 | 62 | −32 | 8.4 |

*Notes*: A negative sign implies net out-migration
[a] = estimated

*Source*: Adapted from Prasaratkul *et al.* (1985), table 3

With regard to fertility, the data presented in Table 2.10 show that the average number of children ever born in the irrigated villages was lower than that in the non-irrigated villages (3.4 and 4.3 respectively). However, the data also show that in one of the irrigated villages (Ban Kok) the average fertility, as well as the fertility within the 20–44 age group, was as high as that in the non-irrigated villages. The authors of the study describe Ban Kok as a deviant case. However, it would have been useful to have more information, such as on the pattern and size of land ownership in the irrigated villages, in order to examine why Ban Kok has a relatively higher level of fertility than the other irrigated villages.

*Table 2.10* Average number of children ever born in irrigated and non-irrigated villages in the Lam Pra Plerng project area (Thailand), 1979

| | *Irrigated villages* | | | | *Non-irrigated villages* | | |
|---|---|---|---|---|---|---|---|
| *Age* | *Suan Hohm* | *Nok Ok* | *Ban Kok* | *All* | *Nong Yai Ray* | *Khun Lakorn* | *All* |
| 15–19 | 1.0 | 0.5 | 0.0 | 0.7 | 0.0 | 0.0 | 0.0 |
| 20–24 | 1.7 | 1.2 | 1.4 | 1.3 | 0.5 | 1.3 | 1.0 |
| 25–29 | 1.9 | 2.3 | 2.4 | 2.2 | 2.6 | 2.2 | 2.4 |
| 30–34 | 2.8 | 2.6 | 4.6 | 3.2 | 4.0 | 3.3 | 3.6 |
| 35–39 | 3.8 | 2.8 | 5.6 | 4.1 | 4.5 | 4.5 | 4.5 |
| 40–44 | 4.4 | 3.4 | 5.6 | 4.5 | 5.8 | 5.4 | 6.5 |
| 45–49 | 4.3 | 3.8 | 6.3 | 5.1 | 7.0 | 9.2 | 7.9 |
| All ages | 3.0 | 2.7 | 4.2 | 3.4 | 4.2 | 4.1 | 4.3 |

*Source*: Adapted from Prasaratkul *et al.* (1985), table 5

From a methodological point of view the choice of experimental and control areas adjacent to one another was, with the advantage of hindsight, rather unfortunate. As the study reports, the building of the dam has resulted in deteriorating conditions for agriculture in at least one of the control villages because of a drastic reduction in its water supply for agricultural activity. The non-irrigated villages have also suffered from the effects of an increase in the cost of living caused mainly by increases in land prices in the irrigated villages. Thus the effects of the irrigation project appear to have been beneficial for some villages and detrimental for others. The net overall effect of the irrigation project could therefore only be determined by assessing the impact of the project in the areas which have benefited from it, along with the impact in the areas which have been adversely affected.

A study by Supapongopichate and Sangsritatankul (1982) was conducted to assess the demographic effects of an artificial lake – the Nam Pong Project in the north-eastern part of Thailand. The study population was divided into three groups: (a) the population around the upstream areas; (b) the population around the reservoir; and (c) the population below the dam. Four villages were selected, using a simple random sampling technique, which provided 374 households to be interviewed.

It was found that out-migration (0.4 persons per household) from the project area was higher than in-migration (0.06 persons per household). With regard to fertility, the study found that the households benefiting from the project indicated a smaller ideal family size than those not benefiting from it. However, in multiple classification analysis when the effect of other independent variables and co-variates was taken into account, the results were the opposite, i.e. those not benefiting from the project desired a smaller family size. The study also concluded that those not benefiting from the project did not want more children.

From the point of view of the methodology, the study thus did not have any control or comparison group. The authors have simply studied the demographic behaviour of the village population in three different locations affected by the dam in different ways. Thus, though one has information on demographic factors pertaining to the population living in these areas, there is no way one can find out what the overall impact of the project has been.

Another study on the demographic impact of irrigation was carried out by Prasith-rathsint *et al.* (1981) in the north-eastern region of Thailand. All the villages in the region were stratified, on the basis of availability and size of irrigation system in the area, into four categories: large, medium, small and none. The size of an irrigation system was defined in terms of total cost of project and the area served by it. From each of the large, medium and small strata, 45 villages were selected with probability proportional to the size of the villages as measured by the number of households; 90 villages without irrigation projects were similarly selected. The total sample included 225 villages scattered throughout the whole region, from which 4,500 households were selected for interview. The fieldwork for the study was conducted between June and November 1980. The analysis was carried out at both the household and the village level, and was confined to the effect of agricultural irrigation systems on fertility.

The results of the regression analysis at the household level presented in Table 2.11 show that only use of electricity, availability of a health station and moderate participation in irrigation were found to be negatively and statistically significantly related to children ever born (CEB) for the sample as a whole. However, it is interesting to note that within each stratum, the variables denoting high, moderate and low participation in irrigation (when significant) are positively related to CEB. The variable related to use of electricity is negatively and significantly related to CEB only in the case of households in villages without an irrigation system. These results do not therefore show any clear indication of the relationship between irrigation and fertility behaviour.

The regression results at the village level presented in Table 2.12 show that the variables related to the percentage of households

*Table 2.11* Regression equations for the determinants of fertility behaviour in north-eastern Thailand, 1980 (household-level analysis) (dependent variable: children ever born)

| *Independent variables* | *Size of irrigation* | | | | |
|---|---|---|---|---|---|
| | *Total* | *Large* | *Medium* | *Small* | *None* |
| Wife's education | −0.005 | −0.015 | 0.035 | 0.014 | −0.006 |
| Male monthly income | −0.000 | −0.000 | −0.000 | −0.000 | — |
| Duration of marriage | 0.200[a] | 0.183[a] | 0.194[a] | 0.197[a] | 0.212[a] |
| Current birth control practice | 0.313[a] | 0.119 | 0.462[a] | 0.384[a] | 0.325[a] |
| Use of electricity | −0.283[a] | −0.167 | −0.275 | −0.240 | −0.329[a] |
| Use of bus and minibus | 0.114 | 0.105 | −0.147 | −0.112 | 0.229 |
| Health station | −0.192[a] | 0.105 | −0.014 | −0.433[a] | −0.274[a] |
| Availability of village school | −0.086 | 0.126 | −0.172 | −0.071 | −0.212 |
| Membership of social organization | 0.074 | 0.231[a] | 0.105 | −0.086 | 0.044 |
| High participation in irrigation | −0.072 | 0.281 | 0.436[a] | −0.449 | — |
| Moderate participation in irrigation | −0.081[a] | 0.054 | 0.119 | −0.058 | — |
| Low participation in irrigation | 0.032 | 0.239[a] | 0.147[a] | 0.000 | — |
| Head of household's participation | −0.195 | −0.163 | −0.043 | −1.085 | 0.212 |
| Constant | 1.317 | 1.017 | 0.894 | 2.542 | 0.833 |
| N | 3311 | 643 | 628 | 690 | 1350 |
| Mean of the dependent variable | 4.6 | 4.4 | 4.5 | 4.7 | 4.6 |

*Notes*: [a] Significant at 5 per cent or less
$R^{2}$ not reported in the study

*Source*: Prasith-rathsint *et al.* (1981), table 5.26

engaged in farming and the size of irrigation system are negatively related to fertility for all types of villages considered together. However, perhaps surprisingly, the percentage of households engaged in farming is significantly and negatively related to fertility only in the case of non-irrigated villages and is not significant in the case of villages with medium or large irrigation systems. (Regression results for villages with small irrigation systems have not been reported.) In the case of villages with medium-sized irrigation systems, the availability of electricity is the only variable which is found to be negatively and statistically significantly related to fertility, though it has a

*Table 2.12* Regression equations for the determinants of fertility behaviour in north-eastern Thailand, 1980 (village-level analysis) (dependent variable: children ever born)

| *Independent variables* | *Size of irrigation* | | | | |
|---|---|---|---|---|---|
| | *Total* | *Large* | *Medium* | *Small* | *None* |
| Percentage farming | –0.1816059[a] | 0.04187745 | –0.00000103 | — | –0.2335575[a] |
| Size of irrigation | –3.544645[a] | — | — | — | — |
| Electricity | –0.053349 | 0.02749120 | –0.00043976[a] | — | –0.1433500 |
| School | –0.906610 | –0.2302008 | –0.02858899 | — | –10.22531 |
| Health station | 0.031640 | –0.04118153 | 0.01646023 | — | 0.5933688 |
| Constant | 30.05456 | 1.878459 | 0.00952515 | — | 41.29364 |
| $\bar{R}^2$ | 0.03 | NR | NR | — | NR |
| N | 225 | 45 | 45 | — | 90 |

*Notes*: [a] Significant at 1 per cent
NR = not reported

*Source*: Prasith-rathsint *et al.* (1981), table 5.4

negligible, almost zero, coefficient. None of the other variables, i.e. percentage engaged in farming, presence of school or availability of a health station, was found to be statistically significant.

For villages with large irrigation systems none of the above-mentioned variables, including availability of electricity, was found to be statistically significant. Thus at the village level also, the analysis does not provide any clear picture of the relationship between irrigation and fertility. This may perhaps be due to the lack of variability in the dependent variable (children ever born) among women in the four groups of villages, i.e. villages with large, medium, small and no irrigation system, as shown in Table 2.11.

It would probably have been more illuminating to categorize villages by their duration of use of irrigation had data been available, since it takes time for an irrigation system to influence socio-economic and demographic factors.

## INTEGRATED RURAL DEVELOPMENT PROGRAMMES

An integrated rural development strategy usually involves the implementation of multi-sectoral projects in a defined geographical area. These often include infrastructural development programmes such as irrigation, roads, electrification and other programmes designed to provide services such as agricultural extension, family planning, health and education.

As discussed earlier, the extent to which a particular project can affect demographic factors depends largely on the availability of other related development inputs. The impact will be more comprehensive when a series of interventions are combined so as to achieve greater efficiencies of social and economic reorganization. Looked at from this point of view, integrated rural development programmes are expected to have a synergistic effect (Hull and Hull, 1988).

A study of an integrated population-development programme in Malaysia – the FELDA (Federal Land Development Authority) Programme – was carried out by Fong (1983), to assess the overall performance of the programme. The major objective of FELDA has been the acceleration of socio-economic development through new land development schemes. In addition to land development, each scheme provides services for social development which include maternal and child health, family planning, youth clubs, co-operatives and education.

Twenty-six out of 300 FELDA schemes were randomly selected for the study. In each of the selected schemes two surveys were conducted: (a) *the staff survey*: to gather information on the organizational factors such as the nature and extent of implementation, resources available and the type of services provided; and (b) *the household survey*: to gather information on the extent to which households support and participate in the programme. The household survey was supplemented by statistical records from the FELDA scheme office.

For the household survey a total of 1,641 settler families were selected at random while 262 scheme officials were selected for the staff survey. The surveys were conducted in the first quarter of 1982.

The performance of the FELDA schemes was measured by a number of indicators which included:

1 *family planning*: proportion of eligible married women practising family planning;
2 *health*: proportion of eligible women who made use of post-natal health care;
3 *education*: proportion of 5–6 year-olds attending kindergarten;
4 *economic*: proportion of husbands and wives involved in extra-mural income-generating activities.

The study also used a number of stratification variables to assess programme performance. These included:

1 length of residence;
2 type of crop grown in the community; and
3 age of housewife.

As far as family planning is concerned, the figures presented in Table 2.13 show that 55.2 per cent of the eligible women were practising family planning at the time of the survey, which was higher than the national rural average of 35.5 per cent. With regard to health, 79 per cent of mothers were making use of post-natal health care as compared to 30 per cent for rural Malaysia as a whole. No comparative national figures were available for educational and economic impact variables. The data also indicate that the programme performance variables do not differ significantly when stratified by length of residence, types of crop grown and age of housewife.

Though the study provides interesting results, in the absence of any control group these could be compared only with the overall national figures. It is rather unfortunate that data were not collected in a non-project area for purposes of comparison using 'with and

*Table 2.13* FELDA programme performance (Malaysia), 1982

| | *Programme performance variables (in percentages)* | | | |
|---|---|---|---|---|
| *Stratification variables* | *Women who practise family planning* | *Women who received post-natal health care* | *Children attending kindergarten*[a] | *Households involved in extra-mural income-generating activities* |
| 1. Length of scheme residence: | | | | |
| 0–2 years | 62.0 | 74.8 | 34.6 | 45.2 |
| 3–5 years | 53.4 | 81.0 | 33.8 | 51.9 |
| > 5 years | 53.4 | 78.8 | 32.0 | 42.2 |
| 2. Type of crop grown in community: | | | | |
| Rubber | 55.7 | 81.4 | 29.1 | 53.6 |
| Oil | 54.8 | 76.8 | 36.9 | 41.2 |
| 3. Age of housewife: | | | | |
| < 20 years | 58.5 | 79.1 | – | 46.5 |
| 20–29 years | 57.5 | 78.1 | 33.8 | 45.1 |
| 30–40 years | 54.9 | 76.7 | 34.2 | 43.2 |
| > 40 years | 53.7 | 73.3 | 33.2 | 48.4 |
| All | 55.2 | 78.9 | 33.3 | 46.9 |
| (National average) | (35.5) | (30.0) | | |

*Note*: [a] For children aged between 5 and 6 years

*Source*: Fong (1983), table 5

without' design, or that no base-line data were available to assess programme performance using 'before and after' design.

Herrin (1988) has assessed the demographic impact of the Bicol River Basin Development Project in the Philippines. The project was set up in 1975 and its major focus has been the development of water resources including expanded drainage, flood control, irrigation and the provision of safe drinking-water supply. Other development projects included in the programme are rural electrification, roads and supporting agricultural, health, nutrition and family planning services. The analysis was carried out using data from the first Bicol Multipurpose Survey (BMS) conducted in 1978. The sample consisted of 1,906 households in 100 rural and urban communities located in three provinces of the region. The study estimated three regression equations for the determinants of fertility (children born during the 1973–8 period) and family planning practices (one for users of any method and one for users of modern methods during the period 1973–7) and also three additional equations for the determinants of current fertility preference and current use of family planning practices (one for users of any method and one for users of modern methods). The definitions of the variables used in the estimation of equations related to fertility and family planning practices are as follows:

| *Variables* | *Definition/measurement* |
|---|---|
| *Dependent* | |
| ADDCHILD | Dummy variable (1 = if the woman reported that she wanted additional children at the time of interview; 0 = otherwise). |
| BIRTH 73 | Number of children born during the period 1973–7. |
| CFPUSEA | Dummy variable (1 = if the woman is using any method of contraception at the time of interview; 0 = otherwise). |
| CFPUSEM | Dummy variable (1 = if the woman is using any modern method of contraception at the time of interview, i.e. pill, IUD, sterilization or injection; 0 = otherwise). |
| FPUSEA 73 | Dummy variable (1 = if the woman used any family planning method during the period 1973–7; 0 = otherwise). |
| FPUSEM 73 | Dummy variable (1 = if the woman used any modern method of contraception during the period 1973–7; 0 = otherwise). |
| *Independent* | |
| *Personal characteristics* | |
| AGEWK | Dummy variable (1 = if the woman's age belongs to category K; 0 = otherwise, where K is coded as<br>1 = age 15–24 years<br>2 = age 25–29 years<br>3 = age 30–34 years<br>4 = age 35–39 years<br>5 = age 40–44 years<br>6 = age 45–49 years). |
| AGEM | Age at marriage in completed years. |
| EWK | Dummy variable (1 = if the woman's level of educational attainment belongs to category K; 0 = otherwise, where K is coded as<br>1 = no schooling or finished up to four years of schooling<br>2 = finished 5–7 years of schooling<br>3 = finished 8 years or more of schooling)<br>(EDW) = [EDW (5–7), EDW (8+)]. |

*(continued on next page)*

| Variables | Definition/measurement |
|---|---|
| PLIVINGCHILD 73 | Number of surviving children prior to 1973, dummy variable (1 = if in category K; 0 = otherwise, where K is coded as<br>1 = 0–2 living children prior to 1973<br>2 = 3–4 living children prior to 1973<br>3 = 5–6 living children prior to 1973<br>4 = 7+ living children prior to 1973).<br>(PLIVINGCHILD) = [PLIVINGCHILD (3–4), PLIVINGCHILD (5–6). PLIVINGCHILD (7+)]. |
| HOUSE | Dummy variable (1 = if the house is made of light construction materials; 0 = otherwise). |
| OWNHOUSE | Dummy variable (1 = if the household owns the house; 0 = otherwise). |
| OWNLAND | Dummy variable (1 = if the household owns agricultural land; 0 = otherwise). |
| *Locational characteristics* | |
| LOC | Location variable (1 = if the community is located in K; 0 = otherwise, where K is coded as<br>CITY = urban (city)<br>POBLACION = municipal *población*<br>RURAL = village).<br>(LOC) = MUN POBLACION, RURAL |
| PROV | Province (1 = if in province K; 0 = otherwise, where K is coded as<br>1 = Sorsogor<br>2 = Albay<br>3 = Camarines Sur).<br>(PROV) = (ALBAY, CAM SUR) |
| RESBGY | Length of residence in community (1 = if household head has resided in community for 5 years or more; 0 = otherwise). |
| *Rural development activities* | |
| AELEC | Dummy variable (1 = if the community is electrified; 0 = otherwise). |
| IRRIG | Dummy variable (1 = if the community has irrigation facilities; 0 = otherwise). |
| TRAVELPOB | Travel time in minutes from the village to the municipal *población*. |

The results of the regression analysis presented in Table 2.14 show that, controlling for other factors, rural electrification and irrigation are significantly related to lower fertility during the period 1973–7. With respect to the use of family planning during the same period, irrigation and travel time to the *población* (town centre), which proxies for the effect of rural road development activities, were found to have significant effects.

*Table 2.14* Estimated regression equations for the determinants of fertility and family planning practices in Bicol River Basin (the Philippines), 1973–7 (t-values are given in brackets)

| *Variables* | *Mean (standard deviation)* | *BIRTH 73* | *FPUSEA 73* | *FPUSEM 73* |
|---|---|---|---|---|
| AGEW73 (25–29) | 0.241 | −0.371[a] | −0.056 | 0.057 |
| | (0.428) | (−3.541) | (−1.039) | (1.243) |
| AGEW73 (30–34) | 0.215 | −0.704[a] | −0.200[a] | −0.059 |
| | (0.411) | (−5.779) | (−3.197) | (−1.122) |
| AGEW73 (35–39) | 0.205 | −1.260[a] | −0.227[a] | −0.135[b] |
| | (0.404) | (−9.418) | (−3.304) | (−2.332) |
| AGEW73 (40–44) | 0.201 | −2.046[a] | −0.468[a] | −0.255[a] |
| | (0.401) | (−14.346) | (−6.402) | (−4.137) |
| EDW (5–7) | 0.485 | 0.002 | 0.109[a] | 0.049 |
| | (0.500) | (0.035) | (3.019) | (1.595) |
| EDW (8+) | 0.232 | −0.334[a] | 0.269[a] | 0.120[a] |
| | (0.423) | (−3.678) | (5.764) | (3.044) |
| AGEM | 20.238 | 0.040[a] | 0.010[b] | 0.005 |
| | (4.210) | (4.778) | (2.327) | (1.414) |
| OWNHOUSE | 0.920 | 0.243[b] | 0.005 | −0.040 |
| | (0.271) | (2.188) | (0.091) | (−0.831) |
| OWNLAND | 0.126 | −0.125 | −0.033 | 0.013 |
| | (0.332) | (−1.383) | (−0.708) | (0.324) |
| MUN POBLACION | 0.156 | −0.193[c] | −0.027 | 0.064 |
| | (0.363) | (−1.752) | (−0.473) | (1.338) |
| RURAL | 0.705 | 0.032 | 0.034 | 0.107[a] |
| | (0.456) | (0.347) | (0.708) | (2.651) |
| RURAL × TRAVELPOB | 45.221 | −0.0002 | −0.001[b] | −0.001[a] |
| | (70.745) | (−0.315) | (−2.136) | (−2.791) |
| RESBGY (5+) | 0.852 | −0.109 | 0.011 | 0.050 |
| | (0.345) | (−1.262) | (0.239) | (1.347) |
| CAM SUR | 0.599 | 0.088 | 0.032 | −0.024 |
| | (0.490) | (0.992) | (0.696) | (−0.626) |
| ALBAY | 0.255 | −0.010 | −0.102[b] | −0.070[c] |
| | (0.436) | (−0.098) | (−1.997) | (−1.633) |
| PLIV CHILD 73 (3–4) | 0.267 | 0.011 | 0.102[b] | 0.029 |
| | (0.443) | (0.120) | (2.243) | (0.752) |
| PLIV CHILD 73 (5–6) | 0.225 | 0.226[b] | 0.159[a] | 0.078[c] |
| | (0.418) | (2.084) | (2.851) | (1.656) |
| PLIV CHILD 73 (7+) | 0.226 | 0.509[a] | 0.184[a] | 0.122[b] |
| | (0.419) | (4.062) | (2.855) | (2.255) |
| AELEC | 0.391 | −0.153[b] | −0.021 | −0.005 |
| | (0.488) | (−2.220) | (−0.595) | (−0.161) |
| IRRIG | 0.565 | −0.134[b] | 0.058[c] | 0.025 |
| | (0.498) | (−2.260) | (1.899) | (0.983) |
| HOUSE | 0.584 | 0.177[a] | −0.140[a] | −0.056[b] |
| | (0.493) | (2.747) | (−4.247) | (−1.994) |
| Constant | | 1.313 | 0.268 | 0.056 |

*(continued on next page)*

*Table 2.14* – continued

| *Variables* | *Mean (standard deviation)* | *BIRTH 73* | *FPUSEA 73* | *FPUSEM 73* |
|---|---|---|---|---|
| $\bar{R}^2$ | | 0.323 | 0.146 | 0.069 |
| N | | 981 | 981 | 981 |
| Mean of the dependent variable (standard deviation) | | 0.431 (0.495) | 0.431 (0.495) | 0.200 (0.400) |

*Notes*: [a] Significant at 1 per cent
[b] Significant at 5 per cent
[c] Significant at 10 per cent
*Source*: Herrin (1988), table 14

Table 2.15 shows the results of regression equations for the determinants of current fertility preferences and current use of contraceptive methods. The data relate to the time of interview when women were asked whether or not they wanted additional children, and whether or not they were practising a specific method of contraception. The results show that none of the three rural development factors (electrification, irrigation and travel to *población*) is significantly related to current fertility preferences. All factors, however, are significantly related to current family planning use (any method) and in the hypothesized direction.

But since no 'control' or 'comparison' groups were used in the study, it is not clear whether and to what extent the changes in fertility and family planning behaviour observed in this region are different from areas where either no such programmes have been implemented, or they have been implemented only partially. Moreover, since the project was set up only in 1975, it is doubtful whether it would have made much impact on fertility during the period 1973–8. It is rather unfortunate that, although a follow-up survey was completed in 1983 providing an opportunity for inter-temporal comparison of the project's demographic effects (Herrin, 1988, p. 92), no use has so far been made of the data from this survey.

A study in Thailand was conducted to assess the demographic impact of Taepa Self-help Settlement (Prasith-rathsint, 1987). Self-help settlements were regarded by the Thai authorities as integrated rural development projects. A number of government agencies were involved in providing technical and other assistance including assistance in physical planning, provision of social services (education, health, etc.) and integrated agricultural programmes covering

*Table 2.15* Estimated regression equations for the determinants of current fertility preferences and current contraceptive practices in Bicol River Basin (the Philippines), 1973–7 (t-values are given in brackets)

| *Variables* | *Mean (standard deviation)* | *ADDCHILD* | *CFPUSEA* | *CFPUSEM* |
|---|---|---|---|---|
| AGEW (25–29) | 0.135 | −0.031 | −0.126 | −0.086 |
| | (0.341) | (−0.441) | (−1.400) | (−1.595) |
| AGEW (30–34) | 0.233 | −0.094 | −0.045 | 0.003 |
| | (0.423) | (−1.355) | (−0.501) | (0.053) |
| AGEW (35–39) | 0.209 | −0.104 | −0.109 | −0.044 |
| | (0.407) | (−1.434) | (−1.162) | (0.778) |
| AGEW (40–44) | 0.199 | −0.150[b] | −0.177[c] | −0.072 |
| | (0.399) | (−2.021) | (−1.848) | (1.253) |
| AGEW (45–49) | 0.195 | −0.173[b] | −0.307[a] | −0.101[c] |
| | (0.396) | (−2.300) | (−3.151) | (−1.731) |
| EDW (5–7) | 0.493 | −0.026 | 0.047 | 0.012 |
| | (0.500) | (−0.985) | (1.397) | (0.594) |
| EDW (8+) | 0.229 | 0.023 | 0.181[a] | 0.057[b] |
| | (0.421) | (0.678) | (4.147) | (2.191) |
| AGEM | 20.119 | 0.002 | −0.0001 | −0.002 |
| | (4.209) | (0.569) | (−0.022) | (−0.800) |
| OWNHOUSE | 0.922 | 0.014 | −0.007 | 0.027 |
| | (0.269) | (0.335) | (−0.127) | (0.838) |
| OWNLAND | 0.129 | −0.019 | −0.004 | −0.034 |
| | (0.335) | (−0.579) | (−0.105) | (−1.340) |
| MUN POBLACION | 0.155 | 0.022 | 0.018 | 0.032 |
| | (0.362) | (0.546) | (0.330) | (1.013) |
| RURAL | 0.708 | 0.052 | 0.044 | 0.013 |
| | (0.455) | (1.492) | (0.987) | (0.495) |
| RURAL × TRAVELPOB | 45.729 | 0.0001 | −0.0004[c] | −0.0002 |
| | (70.568) | (0.525) | (−1.755) | (−1.523) |
| RESBGY (5+) | 0.857 | −0.044 | 0.045 | 0.031 |
| | (0.351) | (−1.389) | (1.105) | (1.291) |
| CAM SUR | 0.601 | 0.046 | 0.013 | −0.064[b] |
| | (0.490) | (1.415) | (0.305) | (−2.502) |
| ALBAY | 0.254 | −0.026 | −0.072 | −0.064[b] |
| | (0.436) | (−0.714) | (−1.501) | (−2.241) |
| LIVING CHILD (3–4) | 0.279 | −0.284[a] | 0.140[a] | 0.025 |
| | (0.449) | (−7.250) | (2.759) | (0.826) |
| LIVING CHILD (5–6) | 0.287 | −0.378[a] | 0.102[c] | 0.003 |
| | (0.452) | (−9.084) | (1.889) | (0.092) |
| LIVING CHILD (7+) | 0.329 | −0.427[a] | 0.092 | −0.026 |
| | (0.470) | (−9.445) | (1.576) | (−0.745) |
| AELEC | 0.386 | 0.015 | 0.072[b] | 0.024 |
| | (0.487) | (0.576) | (2.180) | (1.214) |
| IRRIG | 0.550 | 0.033 | 0.029[c] | 0.008 |
| | (0.498) | (1.482) | (1.936) | (0.453) |
| HOUSE | 0.591 | −0.007 | 0.031[c] | 0.016 |
| | (0.492) | (−0.292) | (−1.894) | (0.885) |

*(continued on next page)*

*Table 2.15* — continued

| *Variables* | *Mean (standard deviation)* | *ADDCHILD* | *CFPUSEA* | *CFPUSEM* |
|---|---|---|---|---|
| Constant | | 0.518 | 0.215 | 0.128 |
| $R^2$ | | 0.181 | 0.090 | 0.045 |
| N | | 1,011 | 1,011 | 1,011 |
| Mean of the dependent variable | | 0.163 | 0.292 | 0.076 |
| (standard deviation) | | (0.370) | (0.485) | (0.265) |

*Notes*: [a] Significant at 1 per cent
[b] Significant at 5 per cent
[c] Significant at 10 per cent

*Source*: Herrin (1988), table 15

production, marketing, loans, agricultural extension services and promotion of collective and co-operative organizations. The study was based on data collected from 650 couples (one half living in the settlement and the other half in the non-settlement villages) during the middle of 1980. Since the Taepa settlement was established in 1955, the survey covered only married women aged 25 years and under at the time of the interview in both settlement and non-settlement villages and those who were single when they settled in the Taepa settlement.

The results of multiple classification analysis presented in Table 2.16 show that living in the Taepa settlement does have a negative effect on fertility. The analysis also shows that those born in the settlement have the lowest level of fertility, followed by those who came in single and those who live outside.

In the absence of any base-line information, the study used a one-shot cross-sectional sample survey, comparing two dissimilar groups and also attempting to add a temporal dimension by including information on place of birth, length of stay in the community and so on. However, the fact that the control villages were too near the experimental area may have influenced the findings.

A study by Khuda (1985) examined the impact of integrated rural development on demographic changes. It was based on data from the village of Sreebollobpur in Comilla-Kotwali Thana (subdivision) in Bangladesh. Comilla-Kotwali is one of the *thanas* in which the Bangladesh Academy for Rural Development (BARD) initiated a programme of integrated rural development in the early 1960s. The BARD programme involved local farmers in broad-based integrated

*Table 2.16* Multiple classification analysis of number of children ever born to married women 25 years and under and to those who came single to live in Taepa settlement (Thailand), 1980 (grand mean = 1.48)

| *Variables and subgroup* | *Sample* | *Unadjusted* | | *Adjusted for independent variables* | | *Adjusted for independent variables and co-variates* | |
|---|---|---|---|---|---|---|---|
| | | *Deviation* | *Beta* | *Deviation* | *Beta* | *Deviation* | *Beta* |
| Type of population | | | | | | | |
| outside self-help community | 47 | 0.14 | | 0.11 | | 0.11 | |
| in self-help community: | | | | | | | |
| born in community | 17 | −0.34 | | −0.26 | | −0.22 | |
| came in as single | 17 | −0.05 | | −0.04 | | −0.10 | |
| came in as married | — | | | — | | — | |
| | | | 0.18 | | 0.14 | | 0.13 |
| Education level | | | | | | | |
| under primary | 4 | −0.08 | | −0.64 | | −0.66 | |
| primary | 52 | 0.17 | | 0.14 | | 0.17 | |
| above primary | 25 | −0.34 | | −0.18 | | −0.24 | |
| | | | 0.22 | | 0.19 | | 0.22 |
| Age at first marriage | 59 | 0.18 | | 0.08 | | 0.07 | |
| under 20 years | 59 | 0.18 | | 0.08 | | 0.07 | |
| 20–24 years | 21 | −0.63 | | −0.33 | | −0.29 | |
| 25–29 years | — | — | | — | | — | |
| over 30 years | — | — | | — | | — | |

| | | | | | | | |
|---|---|---|---|---|---|---|---|
| Duration of marriage | | | | | | | |
| under 4 years | 51 | −0.44 | | −0.34 | | −0.32 | |
| 5–9 years | 30 | 0.75 | | 0.58 | | 0.54 | |
| 10–14 years | — | — | | — | | — | |
| 15–19 years | — | — | | — | | — | |
| over 20 years | — | — | | — | | — | |
| | | | 0.54 | | 0.41 | | 0.38 |
| Women's occupation | | | | | | | |
| agriculture | 50 | −0.00 | | −0.07 | | −0.12 | |
| non-agriculture | 31 | 0.00 | | 0.11 | | 0.19 | |
| | | | 0.00 | | 0.08 | | 0.15 |
| $\bar{R}^2$ | | | | | 0.386 | | 0.439 |

*Source*: Prasith-rasint (1987), table 2

rural development activities which included agricultural extension, credit co-operatives, introduction of improved varieties of seeds and supplementary inputs such as irrigation, fertilizers and pesticides, development of rural infrastructure, education, women's involvement in productive activities and family planning.

The study used a combination of sample survey and anthropological approaches to data collection. The field work was carried out from October 1979 to April 1981. Information on landholding, labour force participation, agricultural production and so on, was collected from all households in the village and information on family planning and fertility was collected from more than 50 per cent of households.

*Table 2.17* Mean number of children ever born to ever-married women in Bangladesh and Sreebollobpur village by age group

| | *Bangladesh* | | | | *Sreebollobpur 1979–81* |
|---|---|---|---|---|---|
| *Age* | *Census 1961* | *DSEP*[a] *1961* | *Census 1974* | *Bangladesh Fertility Survey 1975–6* | *(N)* |
| 15–19 | 0.77 | 0.63 | 0.67 | 0.85 | 0.92 (55) |
| 20–24 | 2.24 | 2.34 | 1.92 | 2.45 | 1.90 (54) |
| 25–29 | 3.51 | 3.89 | 3.29 | 4.24 | 4.04 (54) |
| 30–34 | 4.64 | 5.12 | 4.59 | 5.71 | 4.19 (31) |
| 35–39 | 5.24 | 5.83 | 5.53 | 6.71 | 5.77 (32) |
| 40–44 | 5.49 | 6.08 | 5.83 | 7.10 | 6.00 (29) |
| 45–49 | 5.74 | 6.27 | 6.01 | 6.73 | 6.52 (27) |

*Note*: [a] Demographic Survey of East Pakistan

*Source*: Khuda (1985), table 6

The data presented in Table 2.17 do not provide any clear evidence of the impact of an integrated rural development programme on fertility in Sreebollobpur village, though from Table 2.18 one gets the impression that the fertility rate is lower than average in this particular village.

This study once again underlines the problem of not having baseline data. The author has used national-level data for comparison purposes to assess the programme's impact on fertility. However, the national surveys were not carried out near the time this particular study was undertaken. In fact, surveys used for comparison were carried out in 1961, which is about the same time as the BARD project started.

Moreover, the fact that the fertility rate is lower (Table 2.18) in the experimental village than in the country as a whole does not indicate whether the fertility rate has declined as a result of the programme or whether the village had particular characteristics favouring lower fertility to begin with. To be able to judge the programme's impact on fertility it is thus necessary to know the level of fertility in the village before the programme was implemented.

*Table 2.18* Age-specific marital fertility rates and associated indicators of fertility in Bangladesh and Sreebollobpur village

| *Age* | *DSEP 1960–1*[a] | *Bangladesh Fertility Survey 1975–6* | *Matlab 1979*[b] | *Kotwali 1966–7* | *Sreebollobpur 1979–81 (N)* |
|---|---|---|---|---|---|
| 15–19 | 234 | 310 | 130 | 248 | 432 (37) |
| 20–24 | 337 | 321 | 297 | 279 | 213 (47) |
| 25–29 | 280 | 266 | 302 | 242 | 250 (48) |
| 30–34 | 258 | 229 | 245 | 199 | 172 (29) |
| 35–39 | 161 | 157 | 154 | 126 | 71 (28) |
| 40–44 | 34 | 73 | 40 | 62 | 0 (24) |
| 45–49 | 18 | 27 | 9 | – | 0 (19) |
| TFR | 6.58 | 6.34 | 5.88 | 6.21 | 4.53 |
| CBR | 41 | 46.9 | 35 | n.a. | 30.7 |

*Notes*: [a] Births per 1,000 ever-married women
[b] Age-specific fertility rate
TFR = total fertility rate
CBR = crude birth rate
n.a. = not available

*Source*: Adapted from Khuda (1985), table 5

## LAND REFORM AND AGRICULTURAL RESETTLEMENT SCHEMES

In assessing the impact of land reform on demographic behaviour, at least two dimensions of landholding are important. One is the *size of landholding* to which a household has access for cultivation and the second relates to *land ownership*, including all legal and institutional arrangements that specify how land is to be used and how produce from the land is to be distributed (Stokes and Schutjer, 1984).

The size of holdings can influence fertility by altering the value of children's economic contribution. Farm households with access to larger holdings usually have greater labour requirements. They are therefore able to employ profitably more family labour. This in turn encourages continued high fertility.

Land ownership, on the other hand, can have a negative long-term effect on fertility, operating through the income returns to equity and the resulting increase in old-age security. Land ownership can guarantee an income beyond the period during which an individual is able to work and/or manage an agricultural operation. Thus, at least in theory, land ownership should reduce the importance of children as a source of old-age security and contribute to lower fertility.

Available research on size of holdings and fertility is characterized by a number of empirical and conceptual problems. Most studies that include data on size of landholdings and fertility are not primarily directed at estimating or understanding this relationship but use land as a measure of socio-economic status, wealth or involvement in the agricultural sector. The interpretation given to those findings that do emerge is therefore largely ad hoc in nature. Conceptual questions may be raised as to the causal direction of the effect – that is, whether those with larger holdings have higher fertility, or those with high fertility are able to acquire more land. Most observers agree that the evidence on the relationship between land ownership and fertility is ambiguous since research studies which relate landholdings to fertility do not generally distinguish between the size of holdings to which the household has access for cultivation and the amount of land that is owned.

However, since larger holdings are associated with higher fertility, it is believed that land redistribution programmes are likely to result in increased fertility, at least in the short run. First, as already mentioned, families who experience an increase in the size of their holdings can more profitably utilize child labour. Second, the higher incomes associated with larger holdings can be expected to increase natural fertility and the survival prospects of children.

This interpretation is relevant as regards the short-term response of fertility to land redistribution programmes but is not likely to hold over the long run, particularly if rising incomes alter educational expectations for children. Moreover, redistribution programmes rarely take place in the absence of some tenurial reforms. If land redistribution programmes are accompanied by changes in land ownership, the pro-natalist impact of the redistribution programme may be partially or totally offset. In short, the negative effects on fertility of land ownership may be expected to counteract any increase in the number of surviving children due to redistribution.

The effect of land reform on migration can also be mixed and short-run effects may differ from those in the long run. Land reform, in general, can help to increase household production and income and can thereby slow down rural–urban migration. And if land reform results in larger landholding rather than increased ownership, in the short run increased demand for family labour may also have the effect of slowing down out-migration. However, as mentioned above, since larger landholdings can encourage higher fertility, this may cause population density on land to increase in the long run, causing, in turn, out-migration from the area.

At the community level, added prosperity as a result of land reform can lead to many social changes with modernizing effects, like the opening up of schools and health centres, and access to mass media and recreational activities. These changes can reduce fertility and perhaps migration in the short run. In the long run, however, these modernizing effects may lead to more out-migration because of changes in the attitudes and values of the younger generation.

Land reforms which subdivide large holdings often result in the creation of many small, labour-intensive farms, which helps to increase income and labour utilization in the community. This is expected to reduce out-migration. On the other hand, the subdivision of large holdings may reduce the demand for hired agricultural labour, causing out-migration to increase. Also, increased income and security, as a result of land reform, may provide the farmers with the incentive and ability to mechanize agriculture, thereby further reducing the demand for hired labour and prompting the redundant workers to migrate. The impact of land reform on demographic behaviour is therefore difficult to judge *a priori*. Moreover, the impact of land redistribution or tenure reform will depend on the specific geographic, cultural, socio-economic, institutional and temporal context in which these reforms are carried out.

Land colonization is a special case of land reform which involves resettling rural populations in frontier or less populated areas. The above analysis can equally be applied to study the effects of land colonization programmes on fertility and migration.

Seligson (1979) undertook a study of the effect of land reform on fertility in Costa Rica. A preliminary analysis was first carried out to examine the relationship between land ownership and fertility using data gathered in a survey of 459 peasant households in 1973. The results of this analysis, presented in Table 2.19, show that landed peasants have larger families than landless ones.

*Table 2.19* Analysis of variance and multiple classification analysis of family size among landed and landless peasant households in Costa Rica, 1973

| *Analysis of variance* | *Sum of squares* | *DF* | *F* | *Sig. of F* |
|---|---|---|---|---|
| Main effect | | | | |
| land tenure status | 281.5 | 1 | 25.7 | .001 |
| Co-variates | | | | |
| (age, income, education) | 1577.4 | 3 | 48.0 | .001 |
| Explained | 1858.9 | 4 | | |
| Residual | 4973.6 | 454 | 42.4 | .001 |
| Total | 6832.6 | 458 | | |

Multiple classification analysis
Grand mean = 5.52 (total number of children)
Multiple R = 0.52

| *Type* | *N* | *Unadjusted deviation from mean* | *Adjusted deviation from mean* | *Adjusted mean* |
|---|---|---|---|---|
| Landed[a] | 274 | 0.64 | 0.43 | 5.95 |
| Landless | 185 | −0.95 | −0.61 | 4.91 |
| | 459 | | | |

*Note*: [a] Grouped in the landed category are landowners (both titled and untitled), squatters, renters and share-croppers. The landless peasants include steady plantation and non-plantation workers, day labourers and migrant workers

*Source*: Seligson (1979), table 1

The average family size of the landed peasants is 5.95 as compared with 4.91 for landless peasants – a difference that is statistically significant at the 0.1 per cent level. The difference in family size is not due to differences in age, income or education between the two groups as the effect of these variables has been adjusted by co-variance analysis.

To assess the impact of land reform the study used data from a survey carried out in 1976 consisting of 527 male heads of peasant households who were beneficiaries of the land reform programme and 422 male heads of households who were non-beneficiaries. The latter group included persons from both rural and urban areas (for details of survey design, see Seligson, 1979). The measure used to compare the fertility behaviour of the two groups was the ideal family size. This was used as a proxy for fertility behaviour for two reasons. First, it was argued that the ideal family size reflects the *attitude* of the peasant households towards fertility; second, it was argued that in

Costa Rica there has been a close relationship between decline in ideal family size and decline in fertility.

It was found that the ideal family size among landholding households was greater than that among landless households. The results of the analysis of variance performed with the two samples combined, although not presented here, show that after controlling for age, education and income the ideal family size among the landholding households was on average 4.3 as against 2.8 among landless households.

To investigate further whether or not communal ownership of land reduces fertility among the participants more than individual ownership can, the study sampled 226 communal and 527 individual owners. The results of the analysis of variance performed on the ideal family size presented in Table 2.20 show that there is a significant difference (at less than 1 per cent level) in ideal family size between the two groups. When adjusted for co-variates such as age, income, education, total number of children and 'banana zone',[1] the difference in ideal family size becomes slightly larger. The adjusted ideal family size of the communal land owners was found to be less than that of the individual landowners (3.7 and 4.3 respectively). The analysis therefore shows that land redistribution from communal ownership to individual ownership does indeed encourage people to want larger families and hence is pro-natalist.

A number of methodological points can, however, be raised in connection with the analysis and research design used by Seligson. First, the use of actual family size as a measure of fertility in the 1973 study is perhaps not ideal. It may be that child mortality is higher among the landless peasants, which results in this group having a lower average family size. Second, since 'landowners' is used as a catch-all term encompassing both titled and untitled owners, squatters, renters and share-croppers, the effect of land ownership has not been clearly distinguished from the effect of landholding on fertility. Third, the comparison group in the 1976 study included individuals from *both rural and urban areas* (Seligson, 1979, p. 54). Since fertility in urban areas is generally lower than that in rural areas, this may have biased the results. Fourth, since most beneficiaries of land reform programmes have participated in them only for part of their adult life (Seligson, 1979, p. 54), the study should perhaps have concentrated on the actual fertility behaviour of the beneficiaries of land reform during the post-participation period, rather than on ideal family size. Finally, the justification for using ideal family size as a proxy for actual fertility behaviour is that in two separate studies carried out in 1964 and 1976 in metropolitan Costa Rica, a positive relationship between decline in ideal family size and in actual family size was observed. But it is not clear whether such a relationship, found for metropolitan areas, would also hold true for the rural areas.

*Table 2.20* Analysis of variance and multiple classification analysis of ideal family size of communal and individual peasant owners in Costa Rica, 1976 sample

| *Analysis of variance* | *Sum of squares* | *DF* | *F* | *Sig. of F* |
|---|---|---|---|---|
| Main effect | | | | |
| reform type | 17.3 | 1 | 6.9 | .009 |
| Co-variates | | | | |
| (age, income[a], education, total children, banana zone) | 101.8 | 5 | 8.0 | .001 |
| Explained | 119.1 | 6 | 7.8 | .001 |
| Residual | 1773.6 | 700 | | |
| Total | 1892.7 | 706 | | |

Multiple classification analysis
Grand mean = 4.10 (ideal family size)
Multiple R = 0.25

| *Type* | *N* | *Unadjusted deviation from mean* | *Adjusted deviation from mean* | *Adjusted mean* |
|---|---|---|---|---|
| Communal | 215 | −0.24 | −0.37 | 3.73 |
| Individual | 492 | 0.10 | 0.16 | 4.26 |
| | 707 | | | |

*Note*: [a] A surrogate for income had to be used since all members of each communal enterprise receive the same wage (usually the minimum daily wage for agricultural workers). An index was created based upon the artefacts present in the home (sewing-machine, radio, television, refrigerator, motor cycle, wrist-watch)

*Source*: Seligson (1979), table 2

Seligson's study thus demonstrates that unless due caution is exercised in defining and selecting appropriate variables based on a clear understanding of the mechanisms through which the impact of a specific development intervention on demographic behaviour is felt, it is difficult to draw any meaningful conclusions from statistical analysis.

A study of the impact of agricultural land reform programmes in two districts in north-eastern Thailand was undertaken by Plainoi *et al.* (1982). The study included three groups of villages: villages located in the land reform areas (reform villages), villages adjacent to the land reform areas (adjacent villages) and villages far from the reform areas but close to the second group of villages (distant villages). All together 667 agricultural families from these villages were interviewed for the study. The proportion of families owning land was the highest (90.3 per cent) in the reform villages followed by the adjacent (72 per cent) and the distant villages (68 per cent).

*Table 2.21* Selected demographic characteristics by village types in north-eastern Thailand, 1980

| | *Village type* | | | |
|---|---|---|---|---|
| *Classification* | *Reform villages* | *Adjacent villages* | *Distant villages* | *All villages* |
| Children ever born: | | | | |
| mean | 3.76 | 3.42 | 3.39 | 3.55 |
| N | (276) | (185) | (204) | (665) |
| Children ever born within the past 3 years | | | | |
| mean | 0.85 | 0.85 | 0.95 | 0.88 |
| N | (276) | (187) | (203) | (666) |
| Preferred family size | | | | |
| mean | 4.19 | 4.01 | 4.16 | 4.13 |
| N | (276) | (187) | (204) | (667) |
| Ideal family size | | | | |
| mean | 3.79 | 3.97 | 4.11 | 3.94 |
| N | (276) | (186) | (204) | (666) |

*Source*: Adapted from Plainoi *et al.* (1982), table 5

The data presented in Table 2.21 show that fertility, as measured by the number of children ever born, was highest in the reform villages (3.76), followed by the adjacent (3.42) and the distant villages (3.39). However, when the number of children born within the past three years, from the date of study, was considered it was found that fertility was slightly higher in the distant villages (0.95) than in the reform and adjacent villages (0.85 for both groups).

The study also examined preferred as well as ideal family size in three groups of villages. The data presented in Table 2.21 show somewhat mixed results.[2] Although the average *preferred* family size is the highest in the reform villages, the average *ideal* family size for these villages is the lowest among the three groups of villages.

On the whole, the differences in fertility and family size preferences among the three groups of villages appear to be very small. One reason for this may be that the investigation was carried out a little too soon – less than six years after land reform was implemented. The full effect of the project on fertility may not yet have worked itself out.

In another study Oey (1981) investigated the fertility behaviour of Javanese transmigrants in two resettlement areas of South Sumatra. The study was based on data collected from 1,000 households in three different locations: the two resettlement areas and one area in rural Java, which was the place of birth for the majority of the transmigrants. Complete marital histories were obtained from women aged 25–44.

The study found that fertility among the transmigrants was higher than that among those who stayed in the place of origin. On average the difference between the two groups was 0.8 children. The study also found that some of the higher fertility was due to selection factors. The process for selecting transmigrants tended to be biased towards women in stable marital unions who had at least one child. None the less, most of the higher fertility took place after resettlement. The pregnancy histories showed that fertility rose sharply in the five-year interval after transmigration, and remained higher thereafter.

The co-variance analysis of the difference in fertility (0.8 children) between transmigrants and stayers attributed the difference to the following factors:

1 difference in age distribution accounted for a difference of 0.03 children;
2 difference in educational experiences accounted for a difference of 0.03 children;
3 difference in age at first marriage accounted for a difference of 0.01 children; and
4 other behavioural factors accounted for the remaining difference of 0.73 children.

The study concluded that environmental factors had greatly influenced the fertility behaviour of transmigrant women. Although the conditions for transmigrants and stayers were similar in many respects, the economic environment was different; in the resettlement areas subsistence farming was much more important, and the allocation of land had perhaps resulted in an increased demand for family labour.

It should, however, be mentioned that at the time this research was undertaken family planning programmes had hardly reached the transmigration area covered by the study and this may have exacerbated the effect of environmental factors on fertility (Hull and Hull, 1988).

A study by Uyanga (1985) assessed the impact of the Cross River Plantations Project in Nigeria on fertility and migration behaviour. The study used 'with and without' design. The experimental group covered eight major government plantations and the control group included 200 randomly sampled villages from non-plantation areas. The data were collected in 1979 and the sample size consisted of 2,116 respondents from the plantations and 1,920 from the non-plantation villages.

According to the study, one of the main reasons for establishing plantations between 1950 and 1960 was to redistribute population within the State, encouraging migration from the densely populated south-western part to the eastern region.

With regard to migration the study found that 88 per cent of the respondents in the plantation areas were in-migrants. Of these, 56 per cent had arrived during the initial phase 1950–60, 26 per cent during 1961–70 and 17 per cent during 1971–9. Nearly 70 per cent of the migrants, the majority of whom were males aged 21–30, came from high density areas where it was difficult for them to make a living.

As regards fertility, the study did not find any significant differences in fertility behaviour between plantation and non-plantation population. As data in Table 2.22 show, the two groups of respondents had almost identical preference with regard to family size.

The data in Table 2.23, however, provides some evidence that women's education has an influence on their fertility behaviour in both the plantation and the non-plantation villages.

It is to be noted that in both the plantation and the non-plantation areas, family planning facilities and services were readily available to those who wanted them, though only 11.2 per cent of the plantation population and 8.3 per cent of the non-plantation population were found to be practising family planning. The plantation women who practised family planning more often used it to control the timing of birth rather than to restrict family size.

The desire of a sizeable section of the population, in both plantation and non-plantation areas in the Cross River region in Nigeria, to have a large family seems to be due to prevailing social conditions

*Table 2.22* Percentage distribution of heads of household in plantation and non-plantation areas by additional children wanted and number of living children in Cross River region (Nigeria), 1979

| *Number of living children* | *Additional children wanted* | | | | | | | | |
|---|---|---|---|---|---|---|---|---|---|
| | *0* | *1–2* | *3* | *4* | *5* | *6+* | *Up to God*[a] | *All* | *(N)* |
| Plantation population | | | | | | | | | |
| 0–2 | 0.0 | 4.2 | 4.7 | 12.0 | 19.0 | 21.2 | 38.7 | 100.0 | 784 |
| 3–4 | 0.5 | 4.1 | 8.4 | 10.2 | 16.2 | 23.1 | 38.1 | 100.0 | 645 |
| 5–6 | 1.7 | 0.9 | 16.8 | 23.2 | 15.4 | 16.2 | 26.0 | 100.0 | 539 |
| >6 | 21.3 | 22.7 | 13.4 | 14.4 | 10.2 | 9.7 | 8.3 | 100.0 | 148 |
| All households | 9.4 | 10.6 | 9.6 | 12.4 | 14.0 | 17.5 | 26.5 | 100.0 | 2116 |
| Non-plantation population | | | | | | | | | |
| 0–2 | 0.0 | 3.3 | 5.0 | 10.7 | 12.2 | 20.1 | 48.7 | 100.0 | 814 |
| 3–4 | 0.9 | 4.0 | 6.6 | 10.0 | 19.6 | 20.6 | 38.3 | 100.0 | 621 |
| 5–6 | 1.9 | 1.3 | 7.9 | 16.3 | 18.2 | 19.6 | 34.8 | 100.0 | 300 |
| >6 | 18.7 | 17.0 | 16.5 | 15.1 | 12.0 | 10.5 | 10.2 | 100.0 | 185 |
| All households | 6.1 | 11.1 | 9.3 | 12.2 | 13.9 | 18.7 | 28.7 | 100.0 | 1920 |

*Note*: [a] Includes 'don't know'

*Source*: Adapted from Uyanga (1985), table 7

*Table 2.23* Mean number of children ever born by age and education of wife for plantation and non-plantation population in Cross River region (Nigeria), 1979

| | *Plantation population* | | | *Non-plantation population* | | |
|---|---|---|---|---|---|---|
| *Age* | *No formal education* | *Primary* | *Secondary/ technical or higher* | *No formal education* | *Primary* | *Secondary/ technical or higher* |
| 15–19 | 1.0 | 1.0 | 0.0 | 0.7 | 0.6 | 0.0 |
| 20–24 | 1.4 | 1.6 | 0.0 | 1.8 | 1.8 | 0.0 |
| 25–29 | 2.7 | 2.7 | 2.5 | 2.8 | 3.2 | 2.6 |
| 30–34 | 3.9 | 3.8 | 3.4 | 4.1 | 4.3 | 3.5 |
| 35–39 | 4.6 | 5.5 | 4.9 | 4.9 | 6.4 | 6.5 |
| 40+ | 5.6 | 5.4 | 5.2 | 6.1 | – | – |
| All | 3.4 | 2.8 | 2.7 | 3.6 | 2.9 | 2.8 |
| Standardized for age | 3.4 | 3.1 | 3.0 | 3.6 | 3.3 | 3.2 |

*Source*: Uyanga (1985), table 12

according to which women's status is related to the number of children they have. Thus it was not very fruitful to look for evidence of a reduction in fertility as a consequence of this particular development intervention.

In another study, Henriques (1985) investigated the demographic impact of two large colonization projects in the State of Rondônia in Brazil. Household surveys were undertaken in the two project areas during June and July 1980. Of the two projects investigated, the one in Ouro Preto was the older settlement; it had high-quality soil and was considered well administered. The second project, Gy-Paraná, established in 1972 two years later than Ouro Preto, had poor-quality soil and was considered the most disorganized of all the colonization projects in Rondônia.

With regard to migration, Henriques' study suggests that attracting migrants to the colonization areas was not a big problem. At one point in time, recognizing the impossibility of absorbing the rural population which was migrating to Rondônia in growing numbers, the Government was forced to adopt measures aimed at discouraging these flows at their origin. At the end of 1977, disincentive pamphlets were distributed in several municipalities in the states from which the majority of migrants to Rondônia came (Henriques, 1985, 1988).

As far as fertility is concerned, the data presented in Table 2.24

*Table 2.24* Age-specific and total fertility rates among married women in Ouro Preto and Gy-Paraná (Brazil), 1975 and 1979–80

| Age group | *1975* | *1979–80* | |
|---|---|---|---|
| | *Ouro Preto and Gy-Paraná* | *Ouro Preto* | *Gy-Paraná* |
| 15–19 | 137 | 158 | 400 |
| 20–24 | 444 | 306 | 396 |
| 25–29 | 356 | 300 | 146 |
| 30–34 | 345 | 295 | 192 |
| 35–39 | 192 | 182 | 121 |
| 40–44 | 102 | 100 | 128 |
| 45–49 | 14 | 28 | 26 |
| TFR[a] | 7.95 | 6.84 | 7.04 |
| TFR (Rondônia, rural) | 6.9[b] | 6.8 | |

*Notes*: [a] TFR = total fertility rate
[b] Figure refers to 1970

*Source*: Adapted from Henriques (1985), table 1

show that fertility was still quite high even though it had declined in the two settlements over the period from 1975 to 1979–80. The total fertility rate for the two projects in 1979–80 was found to be around 7, which is about the same as that of rural Rondônia as a whole, according to the 1980 census. It should be mentioned, however, that Brazilian fertility fell rapidly during the 1970–80 decade in every state except Rondônia.

Since one criterion for selecting settlers for the projects was size of family, applicants with large families being given preference over smaller ones, it is quite natural to observe relatively high levels of fertility. However, the decline in fertility over time, although less rapid than in Brazil as a whole, shows that the project has had some impact in reducing fertility in project areas.

Henriques' (1985) study concluded that the high level of fertility that still existed in the project areas could be due to a combination of large demand for family labour and a high level of infant mortality.

Unfortunately, the study did not report fertility rates for the settlers and the *agregados* separately.[3] The latter group was reported to be in a disadvantageous position in terms of landholding. Since landholding patterns can influence fertility behaviour, separate information on the fertility behaviour of the two groups could have been quite revealing.

## WOMEN'S CO-OPERATIVES AND OTHER INCOME-GENERATING SCHEMES

Women's co-operatives and other income-generating schemes have often been conceptualized to affect fertility by providing women with the opportunities for employment and earnings. However, recent studies investigating the link between women's work and fertility have produced contradictory results. Some studies have suggested that where there is a competition for time between women's income-earning activities and the requirements of child bearing and rearing, there may be some decline in fertility. On the other hand, some studies seem to indicate that poor women often value both a large number of children and income-generating work. Recent evidence suggests that women make complex trade-offs between market work and child care; they reduce their own leisure and rest time, select flexible and sometimes poorly paid occupations and substitute the labour of older (particularly female) children for their own labour, both in child care and income-generating activities. In some situations there may not, however, be a conflict between income-generating and child-rearing activities, particularly if the former can be carried out at home.

But, since poor women's demand for a large family is prompted by their perception of risks and uncertainties associated with widowhood, divorce and old age, children – especially male children – are viewed as insurance. Any programme which can reduce the perceived risk and uncertainties during periods of adversity should therefore help to reduce fertility.

Women's co-operatives, training programmes and income-generating schemes that enhance women's skills and their access to resources, and that bring them under modernizing influences, may help to increase their leverage in household decision-making, particularly as related to ideal family size and the sex composition of the family. These programmes can therefore influence the value placed on children and change perceived risks associated with divorce, widowhood and old age (Bruce, 1985).

In Indonesia, a rural co-operative/income-generating scheme which began in 1980 was evaluated in 1984. The major objectives of the scheme were:

1 to improve the quality of life of rural women who practise family planning;
2 to increase the acceptance, practice and continuation of family planning; and
3 to enhance training and management capabilities of women leaders in the rural areas.

Training courses were conducted and funds for a scheme of revolving loans instituted, whereby women's groups received capital which was paid back and then turned over to other groups. Funds were made available only to villages in which at least 35 per cent of women had accepted family planning. In some areas more funds were allocated to groups with higher proportions using preferred methods of contraception. For the purpose of evaluation, field visits were made to beneficiaries of the revolving loan funds. The evaluation, however, was hampered by data limitations and relied primarily on impressions gained during the field visits.

With regard to the objective of improvement of the quality of life, the team concluded that despite 'insufficient empirical data' it was apparent that some women had been able to increase their income. The increased income was reported to have been generally spent on family food and children's education, which in turn was expected to have contributed towards a higher quality of life. The team also observed that the status of women appeared to have been enhanced through their control of funds and their increased activity in the community.

Increased use of family planning, the second objective of the project, was also documented by the team from existing programme statistics at the provincial level. The team members recognized, however, that they needed detailed data to assess the actual impact of the project and to compare results between project and non-project areas.

The third objective of the project, enhancement of training and management skills of women leaders, was considered to have been met based on observations during field visits. The evaluation also indicated that groups continued income-generating activities after the capital had been moved to another group, an indication of continued impact.

Although the team members concluded that the project had had an impact on women's roles and fertility, they noted that their work was inhibited by the lack of sufficient background data and routine statistics on economic and demographic factors (Hull and Hull, 1988).

Another study in Indonesia evaluated an income-generating project in South Sulawasi, which started in 1980 and concentrated on training a group of 120 women leaders to promote and sustain a series of income-generating activities in their communities. The evaluation team used qualitative techniques, including in-depth interviews, informal conversations and observations of a co-operative meeting. Data on family planning were collected both from the records kept by the trained women leaders and National Family Planning Co-ordinating Board.

The evaluation revealed certain anomalies which make it rather difficult to determine precisely the demographic impact of the income-generating activities. Use of family planning was observed to have risen eleven-fold in the region in which the project was located, but similar increases were also observed in non-project areas.

Since all women selected for the training project were already successful acceptors of family planning, it was not possible to test the direct impact of the training project on the family planning behaviour of the trainees. Moreover, since the apparent increase in use of family planning in the community, as claimed by the trainees and attested in the statistics, was not validated with a survey, one has to be cautious in interpreting the conclusion of this study (Hull and Hull, 1988).

A study undertaken in 1980 in Bangladesh (Muhuri and Rahman, 1982) evaluated the demographic impact of a women's co-operative programme and the family planning component of a men's co-operative programme between July 1975 and April 1980. The study adopted a quasi-experimental design, i.e. a design which included 'before' and 'after' observations in experimental and comparison areas. The experimental area consisted of villages in *thanas* (sub-divisions) where a pilot integrated rural development programme (IRDP) was implemented and the comparison area included non-programme villages of the same pilot *thanas*.

A sample of 31 villages was selected from the following four types of villages in the IRDP pilot *thanas*:

Type I: Villages with women's co-operatives only.
Type II: Villages with men's co-operatives only.
Type III: Villages with both men's and women's co-operatives.
Type IV: Villages with no co-operatives.

Of 31 selected villages, baseline data were available for only 14 villages. The selection of neither the pilot *thanas* nor the programme villages was done randomly. A sample of 1,470 households was then drawn from selected villages. Altogether 1,278 currently married female and 963 male respondents were interviewed in April 1980.

*Table 2.25* Total fertility rate by village type, before and after implementation of co-operative programme in Bangladesh, 1976 and 1980

| *Village type* | *Before 1976* | *After 1980* | *Difference* |
|---|---|---|---|
| Villages with both men's and women's co-operatives | 6.60 | 6.06 | −0.54 |
| Villages with women's co-operatives | 6.60 | 6.72 | +0.12 |
| Villages with men's co-operatives | 6.60 | 6.25 | −0.35 |
| Villages without co-operatives | 6.60 | 5.77 | −0.83 |

*Source*: Adapted from Muhuri and Rahman (1982), table 43

The data presented in Table 2.25 show that at the village level, the maximum decline in fertility had taken place in villages where no co-operative programmes had been implemented. In fact, the data indicate that fertility had risen slightly in villages where women's co-operative programmes were implemented. One has to be a little cautious, however, in interpreting these results since baseline data for 1976 were not available by village types (Muhuri and Rahman, 1982, p. 3).

*Table 2.26* Mean number of children ever born (CEB) to married women in co-operative and non-cooperative households in Bangladesh, 1980

| Age | *Women's co-operative members* | | *Non-members of women's co-operative villages* | | *Wives of men's co-operative members* | | *Wives of non-members of men's co-operative villages* | | *Non-members of both men's and women's co-operative villages* | | *Women in non-co-operative villages* | |
|---|---|---|---|---|---|---|---|---|---|---|---|---|
| | *N* | *CEB* | *N* | *CEB* | *N* | *CEB* | *N* | *CEB* | *N* | *CEB* | *N* | *CEB* |
| 10–14 | 2 | 0 | 1 | 0 | 1 | 0 | 4 | 0 | 6 | 0 | 0 | 0 |
| 15–19 | 13 | 0.5 | 32 | 0.6 | 25 | 1.0 | 51 | 0.7 | 66 | 0.6 | 32 | 0.7 |
| 20–24 | 48 | 2.2 | 15 | 2.1 | 38 | 2.3 | 52 | 2.2 | 49 | 1.9 | 33 | 1.7 |
| 25–29 | 41 | 3.5 | 23 | 3.5 | 44 | 4.2 | 36 | 3.4 | 49 | 3.6 | 37 | 3.7 |
| 30–34 | 47 | 4.9 | 14 | 5.2 | 46 | 5.7 | 31 | 5.4 | 37 | 5.5 | 23 | 5.4 |
| 35–39 | 35 | 6.6 | 18 | 7.6 | 43 | 6.8 | 11 | 5.1 | 29 | 6.6 | 19 | 6.6 |
| 40–44 | 33 | 7.1 | 12 | 7.2 | 23 | 8.0 | 11 | 7.1 | 27 | 7.6 | 23 | 7.9 |
| 45–49 | 22 | 7.1 | 17 | 8.7 | 34 | 7.5 | 14 | 7.4 | 9 | 7.9 | 9 | 8.0 |
| All | 241 | 4.6 | 132 | 4.3 | 254 | 5.1 | 210 | 3.2 | 272 | 3.6 | 176 | 3.9 |

*Source*: Adapted from Muhuri and Rahman (1982), table 44

*Table 2.27* Percentage distribution of women according to annual household income in co-operative and non-cooperative villages by membership status in Bangladesh, 1980

| *Annual household income (in taka)* | *Women's co-operative members* | *Wives of men's co-operative members* | *Non-members (females) of both men's and women's co-operative villages* | *Non-members (females) of women's co-operative villages* | *Wives of non-members of men's co-operative villages* | *Women in non-cooperative villages* |
|---|---|---|---|---|---|---|
| | *(N = 239)* | *(N = 254)* | *(N = 272)* | *(N = 131)* | *(N = 210)* | *(N = 168)* |
| Under 3,000 | 8 | 10 | 13 | 11 | 16 | 14 |
| 3,000–5,999 | 27 | 20 | 27 | 26 | 37 | 33 |
| 6,000–11,999 | 27 | 37 | 28 | 37 | 17 | 29 |
| 12,000–17,999 | 18 | 12 | 14 | 14 | 12 | 12 |
| 18,000–23,999 | 8 | 10 | 8 | 4 | 7 | 7 |
| 24,000 and over | 13 | 12 | 10 | 8 | 12 | 5 |
| Total[a] | 100 | 100 | 100 | 100 | 100 | 100 |
| Median income (in taka) | 9375 | 9253 | 8200 | 8125 | 5786 | 6551 |

*Note*: [a] Totals may not add up to 100 because of rounding

*Source*: Adapted for Muhuri and Rahman (1982), table 7

Household level data on age-specific marital fertility reported in Table 2.26 show that members of women's co-operatives and wives of members of men's co-operatives have, on the whole, higher fertility than females not covered by the co-operative programmes. It is also interesting to note that fertility among wives of men's co-operative members is generally higher than other women in most age groups.

*Table 2.28* Percentage distribution of women according to annual household income in co-operative and non-cooperative villages in Bangladesh, 1980

| *Annual household income (in taka)* | *Women's co-operative villages (N = 228)* | *Men's co-operative villages (N = 325)* | *Both men's and women's co-operative villages (N = 555)* | *Non-cooperative villages (N = 163)* |
|---|---|---|---|---|
| Under 3,000 | 10 | 15 | 11 | 15 |
| 3,000–5,999 | 25 | 31 | 26 | 37 |
| 6,000–11,999 | 35 | 23 | 29 | 27 |
| 12,000–17,999 | 17 | 11 | 14 | 12 |
| 18,000–23,999 | 4 | 6 | 10 | 7 |
| 24,000 and over | 10 | 13 | 11 | 5 |
| Total[a] | 100 | 100 | 100 | 100 |
| Median income (in taka)[b] | 8738 | 7105 | 8798 | 6551 |

*Notes*: [a] Totals may not add up to 100 because of rounding
[b] Median incomes were calculated from ungrouped data

*Source*: Adapted from Muhuri and Rahman (1982), table 8

Both the household and village level income data presented in Tables 2.27 and 2.28 show that both women's co-operative members and wives of men's co-operative members belong on average to better-off sections of the community, which makes the findings a little more surprising.

However, the data on earnings presented in Table 2.29 show that women who are members of co-operatives do not have a large monthly income. Only 21 per cent of the members have monthly earnings above 91 takas and their income constituted about 13 per cent of annual household income. The evidence therefore suggests that co-operative programmes have not yet significantly enhanced women's capacity for earning. In addition, four years is too short a period for the full effects of such programmes to be felt upon income and fertility behaviour.

*Table 2.29* Percentage distribution of women by membership status and average monthly earnings in Bangladesh, 1980

| *Membership status* | *Average monthly earnings (in taka)* | | | | | | |
|---|---|---|---|---|---|---|---|
| | *No earnings* | *up to 30* | *31–60* | *61–90* | *91+* | *Total*[a] | *N* |
| Women's co-operative members | 59 | 10 | 8 | 1 | 21 | 100 | 240 |
| Non-members of women's co-operative villages | 83 | 9 | 3 | 1 | 5 | 100 | 866 |
| Women in non-cooperative villages | 87 | 6 | 2 | 1 | 4 | 100 | 168 |

*Note*: [a]Totals may not add up to 100 because of rounding

*Source*: Adapted from Muhuri and Rahman (1982), table 54

Another study by Rahim and Mannan (1982) has evaluated the impact of a women's vocational and training (WVT) programme on fertility behaviour in Bangladesh. The evaluation was carried out about three years after the implementation of the programme in 1979. The major objective of the WVT programme was to improve rural women's socio-economic conditions by upgrading their skills and to prepare them for acting as agents for social change.

The respondents for the study were randomly selected from three groups:

1 521 ever-married women directly exposed to the programme, of which 257 were currently married;
2 655 currently married women from the programme villages but not directly exposed to the programme;
3 312 currently married women from non-programme villages.

The survey was carried out during September and October 1979. Data presented in Table 2.30 show that the total fertility rate of the women directly exposed to the programme (3.6) was the lowest in comparison to that of indirectly exposed women (7.5) and women in non-programme villages (7.2).

The relatively low fertility among the directly exposed women may not, however, be due to the effect of the programme, particularly since the programme has been in existence only for about three years.

Moreover, as Table 2.30 shows, the directly exposed women are relatively younger than women in other categories. Also, the former group of women are relatively more educated than the latter.

*Table 2.30* Age-specific fertility rates by exposure to women's vocational training programme (WVT) in Bangladesh, 1979

| | *Exposure to WVT programme* | | | | | |
|---|---|---|---|---|---|---|
| *Age* | *N* | *Direct exposure* | *N* | *Indirect exposure* | *N* | *No exposure* |
| 15–19 | 51 | 215.7 | 85 | 218.8 | 38 | 302.3 |
| 20–24 | 82 | 146.3 | 134 | 365.0 | 50 | 260.0 |
| 25–29 | 49 | 122.5 | 124 | 266.0 | 52 | 250.0 |
| 30–34 | 30 | 100.0 | 106 | 207.0 | 55 | 166.7 |
| 35–39 | 16 | 123.0 | 74 | 216.0 | 49 | 180.0 |
| 40–44 | – | – | 90 | 112.0 | 38 | 222.2 |
| 45–49 | – | – | 41 | 24.0 | 21 | 50.0 |
| TFR[a] | – | 3.6 | – | 7.5 | – | 7.2 |

*Note*: [a] TFR = total fertility rate

*Source*: Rahim and Mannan (1982), table 16

*Table 2.31* Percentage distribution of women by education and exposure to WVT Programme in Bangladesh, 1979

| *Education* | *Direct exposure* | *Indirect exposure* | *No exposure* |
|---|---|---|---|
| Illiterate | 46 | 73 | 73 |
| Primary | 26 | 14 | 14 |
| Above primary | 28 | 13 | 13 |
| Total | 100 | 100 | 100 |
| Median years of education | 3.3 | 0.7 | 0.7 |
| N | 255 | 655 | 312 |

*Source*: Adapted from Rahim and Mannan (1982), table 2

Thus it seems that the women who received training belong to a self-selected group of relatively younger women with more education than indirectly exposed women and women in non-programme villages. Hence, though at a first glance the impact of the project on fertility may seem to be yielding the desired result, one cannot be sure of its success or otherwise until the programme is evaluated again after, say, ten years or so, with a control or comparison group having similar age and educational level as that of the project women.

A study in 1980 by Siddiqui *et al.* (1982) evaluated the impact of the mothers' club programme on fertility in Bangladesh. The programme, initiated in 1975, was aimed at setting up mothers' clubs in 40 villages in each of the 19 *thanas* selected. All village women were eligible for membership. Women who were club members received vocational training and instructions in family planning and basic education. Access to contraceptive supplies and recreational facilities were also made available as part of the programme.

The study used multi-stage random sampling procedure to select four *thanas* and then two mothers' clubs from each *thana*. A total of 75 households were randomly selected from each *thana* giving a sample size of 600. Of these 195 respondents were direct participants or members of mothers' clubs and the rest, 471, were indirect participants, i.e. women who were not members of the mothers' clubs but lived in villages where the programme was being implemented. Of the 195 direct participants, however, information relating to only 113 currently married women was used for analysis. The remaining 82 women were unmarried at the time of the survey. In addition, a sample of 248 households was selected from villages where the programme was not in operation. The survey was carried out during January and February 1980.

The data on fertility reported in Table 2.32 show that total marital fertility was the lowest among the direct participants (4.7) compared to that of the indirect participants (7.5) and non-participants (9.5). However, no statistically significant relationship was found between programme exposure and current fertility (number of births during the last one year). The mean number of children ever born for the three groups were 4.1 for direct participants, 4.4 for indirect participants and 4.0 for non-participants.

The reason why the participants had lower total fertility may be that on average the direct participants in the programme were younger (median age 26 years) than the indirect participants (median age 30 years) or non-participants (28 years). The women were also better educated. The literacy rate among the three groups was found to be 60, 33 and 21 per cent respectively. Since the period for which the participants were exposed to the programme is again very short (two to four years) for any impact on fertility to be felt, it is difficult to conclude whether the observed lower fertility among participants is due to the effect of the programme or due to the specific characteristics of the participants such as age and level of education favouring lower fertility.

*Table 2.32* Socio-demographic characteristics of participant and non-participant women in the rural mothers' club programme in Bangladesh, 1980

| *Indicator* | *Participation status* | | |
|---|---|---|---|
| | *Direct participants* | *Indirect participants* | *Non-participants* |
| Total fertility rate | 4.70 | 7.51 | 9.50 |
| Average number of children ever born | 4.1 | 4.4 | 4.0 |
| Median age | 26 | 30 | 28 |
| Literacy rate (per cent) | 60 | 33 | 21 |

Age-specific fertility rates:

| *Age of mother* | *(N)* | *Fertility rate* | *(N)* | *Fertility rate* | *(N)* | *Fertility rate* |
|---|---|---|---|---|---|---|
| Less than 15 | ( 3) | – | ( 6) | – | ( 2) | 0.50 |
| 15–19 | ( 13) | 0.15 | ( 65) | 0.42 | ( 40) | 0.38 |
| 20–24 | ( 19) | 0.21 | ( 81) | 0.26 | ( 50) | 0.22 |
| 25–29 | ( 28) | 0.18 | ( 78) | 0.32 | ( 53) | 0.32 |
| 30–34 | ( 14) | 0.29 | ( 71) | 0.23 | ( 44) | 0.23 |
| 35–39 | ( 19) | 0.11 | ( 85) | 0.21 | ( 25) | 0.22 |
| 40–44 | ( 7) | – | ( 57) | 0.70 | ( 16) | 0.60 |
| 45 and above | ( 10) | – | ( 26) | – | ( 18) | – |
| All women | (113) | 0.94 | (469) | 1.50 | (248) | 1.90 |

*Source*: Adapted from Siddiqui *et al.* (1982), p. 6 and tables VC 12.61–3

## RURAL JOB CREATION PROGRAMMES

The aim of rural job creation programmes has predominantly been to influence migration. However, by generating employment and income for the rural population, these programmes can also affect fertility through changes at both household and community level.

At household level, increased opportunities for employment and earnings can provide household members with the incentive to remain in the rural area, thereby helping to reduce out-migration. Increased household earnings may also generate demand for new goods, better health and education and so on. This, in turn, can affect fertility by changing attitude towards ideal family size. Rural job creation programmes can also lead to attitudinal changes by providing household members with greater opportunities to participate in community activities.

At the community level rural job creation programmes may promote new industrial and business activities in response to new demands generated at the household level. Construction of roads, irrigation and other development projects which may be a part of the rural job creation programme can contribute towards infrastructural development of the community, thereby enhancing the attractiveness of the rural community to potential migrants. Increased income of the community as a whole can promote social and cultural activities which can help to change its attitude towards family size and create demand for family planning.

The extent to which a rural job creation programme can affect demographic factors will of course depend on the nature, scope and size of the programme itself. For example, if income generated by a job creation programme is not large enough, or if the benefits from the programme fail to reach the potential migrants, such programmes will fail to generate the desired results.

A study was undertaken by Rodmanee and Bunnag (1983) to assess the impact of a rural job creation programme (RJCP) in reducing rural–urban migration during the dry season (January–March) in poor provinces of southern Thailand. For this study, 818 household heads were selected from each of the two types of households; those with members employed in the rural job creation programme and those without. The survey was carried out during October and November 1981.

The results of multiple classification analysis, presented in Table 2.33, show that propensity to migrate among farmers with holdings of more than 10 rai, non-participants in the programme and households in richer districts was relatively lower than among landless and marginal farmers, participants in the programme and those belonging to poorer districts. The respondents who were relatively older also had a lower propensity to migrate than the younger ones.

Overall these results show that the rural job creation programme was unable to reduce seasonal out-migration from rural areas. The major reason was perhaps that the programme has had very little impact on employment and income of the participants. The data presented in Table 2.34 show that people who participated in the rural job creation programme were employed for a very short period (7.4 days on average during the dry season) and income earned from the programme was only a small proportion of the total income. Thus the

*Table 2.33* Multiple classification analysis of seasonal out-migration, southern Thailand, 1981

| | | Grand Mean = 0.04 | | | |
|---|---|---|---|---|---|
| *Variable* | *N* | *Unadjusted means* | *Eta* | *Adjusted means* | *Beta* |
| Main occupation | | | 0.11 | | 0.07 |
| non-agricultural | 129 | 0.05 | | 0.03 | |
| agricultural | 689 | −0.01 | | −0.01 | |
| Employment in RJCP | | | 0.04 | | 0.08 |
| not employed | 464 | −0.01 | | −0.01 | |
| employed | 354 | 0.01 | | 0.02 | |
| Size of landholding | | | 0.08 | | 0.06 |
| none | 48 | 0.05 | | 0.02 | |
| 1–10 rai[a] | 372 | 0.01 | | 0.01 | |
| 11–12 rai | 216 | −0.02 | | −0.02 | |
| over 20 rai | 182 | −0.01 | | 0.00 | |
| Type of district | | | 0.05 | | 0.05 |
| rich | 421 | −0.01 | | −0.01 | |
| poor | 397 | 0.01 | | 0.01 | |
| Age of respondents | | | 0.14 | | 0.11 |
| under 30 years | 164 | 0.05 | | 0.04 | |
| 30–34 | 129 | 0.00 | | 0.00 | |
| over 34 | 525 | −0.02 | | −0.01 | |
| $R^2$ | | | | 0.067 | |

*Note*: [a] 1 rai = 0.16 hectare

*Source*: Adapted from Rodmanee and Bunnag (1983), table 12

*Table 2.34* Employment and income of the participants and non-participants of the rural job creation programme, southern Thailand, 1981

| *Employment/income during the dry season (January–March)* | *Average* | *Standard deviation* |
|---|---|---|
| Participant | (N = 314)[a] | |
| Length of employment in the programme (days) | 7.4 | 0.5 |
| Income earned from the programme (baht per day) | 69.0 | 3.2 |
| Length of employment in other work (days) | 58.2 | 1.9 |
| Income earned from other work (baht per day) | 78.2 | 7.0 |
| Non-participants | (N = 328) | |
| Length of employment during the dry season (days) | 81.1 | 3.1 |

*Note*: [a] Data were not reported for the whole sample

*Source*: Adapted from Rodmanee and Bunnag (1983), table 8

rural job creation programme appears to have provided very small financial benefits to those who have participated in it and as a result has failed to deter rural–urban migration. The results of this study therefore suggest that in general there is little point in searching for demographic impacts in situations where the development intervention has not generated sufficient benefits.

## PROMOTION OF SMALL-SCALE INDUSTRY

The effect of promotion of small-scale industries on fertility cannot be determined *a priori*. The increased income from participation in small-scale industries is more likely to reduce fertility, particularly in the long run. But in the short run, these industries may in fact encourage higher fertility. This may happen for two reasons. First, income-earning activities at home may not be in conflict with child-rearing activities. Second, since these industries are usually labour-intensive, increased demand for family labour in the wake of promotion of small-scale industries is likely to increase demand for children.

With regard to migration, the promotion of small-scale household industries is likely to generate additional income and employment at the household level, and can therefore reduce out-migration at least in the short run. By creating employment opportunities for family labour throughout the year, promotion of these industries may also help to deter seasonal out-migration during off-farming seasons. However, if the effect of promotion of small-scale industries on fertility is pro-natal, it may encourage out-migration from rural areas, particularly in the long run, by increasing pressure of population on land.

A study on the effects of small-scale industry on fertility was carried out by Chalamwong (1983) using data collected from 247 households in 22 villages in three provinces in Thailand. Altogether, 261 currently married women in these 247 households were interviewed. The field survey was carried out from March 1980 to February 1981.

The initial examination of data revealed that there exists a non-linear relationship between female labour force participation in small industry and fertility. Number of children ever born increased consistently with women's total working hours, reaching its peak between 1,500 and 2,000 hours per year, and then declined consistently. A similar pattern of relationship occurred when female labour force participation was measured in terms of women's working hours on off-farm activities. Children ever born peaked between 500 to 1,000 working hours per year. However, no significant relationship was found between fertility and number of years engaged in home industry.

The study also used a simultaneous equations model, which treated fertility (children ever born) and women's participation in the labour force as endogenous variables. Five different measures of women's labour force participation were used. The results of the two-stage least squares method reported in Table 2.35 show that women's participation under the first four measures has a negative effect on children ever born. But a contradictory result was obtained when women's participation was measured in terms of years engaged in home industry, which showed that fertility is positively related to female labour force participation. For all five estimated equations, child labour force participation has a small but positive effect on children ever born.

On the basis of the results of the first four regression equations the study concluded that efforts aiming at increasing regular hours in productive work among married women in rural areas should lead to fertility reductions. However, the estimated coefficients of the variable related to women's labour force participation in the first four equations, though they have negative signs, are very small ( - 0.0062 to - 0.0083). The exception is a relatively large positive coefficient of 0.49 related to the same variable when it is measured as 'number of years the wife has been engaged in home industry'. The study seems to discount this result on technical grounds. But, as discussed above, female participation in home industry can, at least in the short run, give rise to increase in demand for children particularly if they can be used as labour in home industry. Since all five equations estimated show that participation of child labour is positively related to fertility (children ever born), one perhaps should not dismiss the result of the last regression equation too quickly.

A study carried out in Bohol province in the Philippines in 1980 (Pernia and Pernia, 1986) attempted to assess the impact of small-scale (employing five to 99 workers) and cottage (employing one to four workers) industries on income, employment, health, labour force participation and fertility. The data for the study were obtained from two linked surveys - one for enterprises and another for households. The enterprise survey could collect information from only 85 enterprises out of 180 planned. Of these 85 enterprises, 31 received assistance from Medium and Small Industries Co-ordinated Action

*Table 2.35* Estimated regression equations for the determinants of fertility, Thailand, 1980–1 (dependent variable: children ever born; t-values are given in brackets)

| *Exogenous variables* | *Model 1* | *Model 2* | *Model 3* | *Model 4* | *Model 5* |
|---|---|---|---|---|---|
| Constant | 4.6137 | 1.8025 | −0.0437 | 0.1337 | 2.4370 |
| Household income | $0.53 \times 10^{-5a}$ | $-0.73 \times 10^{-5a}$ | $-0.90 \times 10^{-5a}$ | | $-0.74 \times 10^{-5a}$ |
| | (5.3) | (7.3) | (9.0) | | (7.4) |
| Wife's labour force participation[1] | −0.0062[a] | −0.0054[a] | −0.0083[a] | −0.0070[a] | 0.4909[a] |
| | (21.38) | (16.87) | (17.44) | (2.42) | (7.4) |
| Children's labour force participation | $0.21 \times 10^{-3a}$ | $0.29 \times 10^{-3a}$ | $0.15 \times 10^{-3a}$ | $0.52 \times 10^{-3a}$ | $0.41 \times 10^{-3a}$ |
| | (3.00) | (3.62) | (1.87) | (4.72) | (3.73) |
| Size of cultivated land | 0.1430 | −0.0544[a] | −0.0196[a] | −0.0067 | −0.0588[a] |
| | (0.35) | (8.67) | (3.92) | (0.75) | (4.74) |
| Quality of land | −2.2382[a] | −2.1381[a] | −0.4147[a] | −0.3662 | 1.7600[a] |
| | (13.08) | (10.58) | (2.78) | (1.18) | (5.5) |
| Land ownership | 0.4381[a] | −0.1719 | −1.4482[a] | 0.1663 | −0.3531[c] |
| | (3.10) | (1.06) | (8.09) | (0.68) | (1.76) |
| Birth control practice | 0.0531 | 0.0841 | 0.0143 | 0.1421 | 0.1021 |
| | (0.48) | (0.64) | (0.12) | (0.75) | (0.57) |
| Age of wife | 0.3577[a] | 0.2944[a] | 0.2237[a] | 0.2038[a] | −0.6281[a] |
| | (32.51) | (25.85) | (23.30) | (13.86) | (14.74) |
| Wife's age at first marriage | −0.6406[a] | −0.0902[a] | −0.0141 | −0.1275[a] | −0.0811[b] |
| | (30.5) | (3.75) | (0.56) | (3.54) | (2.38) |
| Wife's education | −0.4311[a] | −0.0810[c] | 0.2561[a] | −0.1547[c] | −0.1059[c] |
| | (7.72) | (1.64) | (4.05) | (1.64) | (1.64) |
| $R^2$ | 0.8323 | 0.7775 | 0.7843 | 0.5310 | 0.5888 |
| N | 261 | 261 | 261 | 261 | 261 |

*Notes*: [a] Significant at 1 per cent, [b] Significant at 5 per cent, [c] Significant at 10 per cent

[1] Definition for labour force participation used for models 1–5 are, respectively:
(i) hours spent on all economic activities, (ii) hours spent on off-farm work, (iii) hours spent on wage labour, (iv) hours spent in home industry, (v) years spent in home industry

*Source*: Adapted from Chalamwong (1983), tables 8–12

Programme (MASICAP) of the Ministry of Industries and the Small Business Advisory Centre (SBAC),[4] 20 were assisted by other programmes and 34 were unassisted.

The household survey obtained information from 428 households out of 530 planned. The sample consisted of 29 owners' households of MASICAP/SBAC (M/S)-assisted enterprises, 91 owners' households of non-M/S assisted enterprises, 89 hired workers' households of M/S-assisted enterprises, 90 hired workers' households of non-M/S assisted enterprises and 129 households not connected with the small industries programme.

*Table 2.36* Estimated regression equation for the determinants of household income, Bohol (the Philippines), 1980 (t-values are given in brackets)

| *Independent variables* | *Regression coefficients* |
|---|---|
| Constant | −5170.859 |
| Household labour | 1.814[a]<br>(6.781) |
| Education of household head (years) | 688.759[a]<br>(5.268) |
| Age of household head | 53.632<br>(1.296) |
| Owner with M/S | 10991.305[a]<br>(5.214) |
| Owner without M/S | −1851.931<br>(−1.320) |
| Worker with M/S | −2132.907<br>(−1.484) |
| Worker without M/S | −2014.653<br>(−1.401) |
| $\bar{R}^2$ | 0.29 |
| N | 428 |

*Note*: [a] Significant at at least 5 per cent

*Source*: Adapted from Pernia and Pernia (1986), table 6

The study examined the impact of M/S assistance on fertility only indirectly. First, the effect of the M/S programme on income was analysed. The results of the regression analysis presented in Table 2.36 show that M/S assistance had tended to increase incomes, particularly of owners of enterprises. Higher incomes were then found to have a negative effect on fertility after a certain threshold level (4,000 pesos) was reached (see Table 2.37). The results presented in Table 2.37 also show that while the wife's level of education and age at marriage have a negative effect on fertility, the wife's current age is positively related to the number of children ever born.

*Table 2.37* Estimated regression equation for the determinants of fertility (CEB), Bohol (the Philippines), 1980 (t-values are given in brackets)

| *Independent variables* | *Regression coefficients* |
|---|---|
| Constant | 4.449 |
| Below threshold income[1] | 0.000 |
| | (0.467) |
| Above threshold income[2] | −0.000 |
| | (−1.607) |
| Education of wife (years) | 0.247 |
| | (1.579) |
| Education of wife squared | −0.011 |
| | (−1.438) |
| Wife's age 20–24 | 2.133[a] |
| | (2.260) |
| Wife's age 25–29 | 3.127[a] |
| | (3.527) |
| Wife's age 30–34 | 4.413[a] |
| | (5.040) |
| Wife's age 35–39 | 5.625[a] |
| | (6.439) |
| Wife's age 40+ | 7.241[a] |
| | (8.410) |
| Wife's age at marriage | −0.295[a] |
| | (−13.195) |
| $\bar{R}^2$ | 0.529 |
| N | 337 |

*Notes*: [1] Below threshold income = min. (0, annual household income − P 4,000)
[2] Above threshold income = max. (0, annual household income − P 4,000) (Annual household income is in Pesos)
[a] Significant at at least 5 per cent

*Source*: Adapted from Pernia and Pernia (1986), table 7

One has, however, to be a little cautious in interpreting the results of the regression analysis on fertility, since the coefficients of the threshold income variables are negligible and statistically insignificant. Moreover, the dependent variable, children ever born, includes events in the past that cannot possibly be influenced by the current impact of M/S assistance. Measures of more recent fertility, temporally consistent with the onset of M/S assistance, would have been preferable. The findings related to wife's current age, age at marriage and education are perhaps not too surprising.

It would have been more informative if the study had directly estimated the effect of participation of women in M/S assisted industries on fertility.

A study undertaken by Bhattacharya and Hayes (1983) assessed the effect of cottage industry on fertility in a rural area of West Bengal in India. The study was based on a survey of 362 families who were participating in cottage industry. The fieldwork was carried out in 1980. Questions concentrated on each family's fertility experience and on age at first participation of family members in the industry.

From data collected, the study reconstructed to some extent the history of each family's fertility behaviour and participation in cottage industry. From this the study determined two events: (a) at what age a woman first participated in cottage industry, and (b) at what age at least one of her children first participated in the industry. It then compared the fertility of women before and after these two events with the fertility of women in the same age group who did not participate in cottage industry.

*Table 2.38* Mean number of children born to women by work participation of children in cottage industry, West Bengal (India), 1980

| Current age | *Women with at least one child participating* | | | | *Women with no child participating* | |
|---|---|---|---|---|---|---|
| | *Before participation* | *After participation* | *Total* | *N* | *Total* | *N* |
| 20–24 | – | – | – | 0 | 1.65 | 43 |
| 25–29 | 3.60 | 0.10 | 3.70 | 10 | 3.19 | 43 |
| 30–34 | 3.32 | 0.24 | 3.56 | 25 | 3.26 | 47 |
| 35–39 | 4.27 | 0.39 | 4.67 | 33 | 3.79 | 14 |
| 40–44 | 4.54 | 0.34 | 4.89 | 35 | 3.78 | 9 |
| 45–49 | 4.70 | 0.20 | 4.90 | 20 | 3.73 | 11 |

*Source*: Adapted from Bhattacharya and Hayes (1983), table 4

First, the data in Tables 2.38 and 2.39 show that women with children working in cottage industries have higher fertility than women whose children do not work in these industries. Second, women who have at least one child working in the cottage industry had a relatively large number of children compared to other women, even before one of the children first started to work in this industry. Third – and perhaps most interesting of all – women with children working in the cottage industries have fertility as low or even lower than women with no children working, once their children started working. In other words, children's participation in the cottage industry appears to cause a subsequent decline in fertility. The main conclusion of the study was that high fertility caused children's participation in cottage industries. But once their participation occurred, it acted as a break on additional fertility.

*Table 2.39* Births per woman per year by work participation of children in cottage industry, West Bengal (India), 1980

| *Current age of mother* | *Women with at least one child participating* | | *Women with no child participating* |
|---|---|---|---|
| | *Before participation* | *After participation* | |
| 25–29 | 0.30 | 0.06 | 0.18 |
| 30–34 | 0.26 | 0.04 | 0.12 |
| 35–39 | 0.15 | 0.04 | 0.04 |
| 40–49 | 0.15 | 0.03 | 0.02 |

*Source*: Adapted from Bhattacharya and Hayes (1983), table 5

From the point of view of the methodology, this study is an interesting example of finding a way around the difficulties in impact assessment due to lack of appropriate longitudinal data.

However, as the study itself has pointed out, 'the ideal research design would probably have been a panel study', following families through time looking at their work and fertility history in order to examine their fertility behaviour before and after they started participating in cottage industries.

Chapter 3

# Appropriate methodologies for impact assessment

The purpose of this chapter is twofold: first, to review the major shortcomings of the methodologies used by the existing studies assessing demographic impacts discussed in the previous chapter, and, second, to suggest ways of improving existing methodologies and the future directions which research on demographic impacts might take in order to guide policy decisions.

One of the most crucial aspects of demographic impact assessment is the selection of an appropriate *study design*. It is not easy to undertake the classical experimental design, whose major features are (a) the random assignments of study units such as households and villages into experimental and control groups, and (b) collection of data related to periods before and after the specific development intervention. Implementing a quasi-experimental (or 'before' and 'after') design that does not include control groups but involves collection of survey data before project implementation and after project maturation also often takes too much time and resources. Moreover, in the meantime, various contemporary changes begin to contaminate the impact of the project, making it difficult to disentangle the effects of a specific project from the effects of other changes which have taken place in the experimental area.

Very often, then, researchers are left with the option of using non-experimental or 'with' and 'without' design. In this design, random selection of units into experimental and control groups is usually not possible. Hence the comparison group is selected using the evaluator's judgement or on the basis of some prior information as to the approximate pre-project equivalence between two groups in terms of characteristics that might be related to programme outcomes. The advantage of this design is that it can easily be implemented. The major disadvantage, however, is that the observed differences in certain key variables, between the experimental and control areas, may not reflect the net impact of the programme but rather the effects of (a) systematic differences between the two groups prior to the programme, and (b) differential exposure to contemporaneous events other than the programme.

Pekanan's (1982) study on the impact of rural electrification in Thailand reviewed in Chapter 2 provides a clear example of these problems encountered in impact assessment which can sometimes give rise to misleading results. The study, using a 'with and without' design, concluded that rural electrification was the most significant factor related to a decline in fertility in the electrified village as compared to the non-electrified 'control' village. Yet a closer look at the data showed that fertility in the 'experimental' village was lower than that of the 'control' village both before and after the 'experimental' village was electrified. Moreover, the study did not find social conditions any different in the two villages either before or after one of the villages was electrified. Also, only 4.5 per cent of households in the electrified village were found to be using electricity for purposes other than lighting. By the time the study was undertaken the electrified village had had access to electricity for only four years. Since this is too short a period for the effect of electrification to be felt on fertility, one would question the conclusion that the lower fertility in the experimental village was due to electrification.

Another related problem is that of *selection bias* in the study units, which often makes it difficult to derive scientifically valid conclusions. Development projects are usually directed towards target populations which often possess characteristics different from others. For example, projects may be first introduced in more depressed areas where people are in greater need of public intervention. Similarly, richer households in project areas may be the first ones to be connected to an electricity supply. Thus to the extent that there is a selection bias in the choice of areas or households to receive a project, it is not correct to attribute demographic differences to a particular project.

For example, the study in Bangladesh (Rahim and Mannan, 1982) found that fertility among a group of women who received vocational training was lower than that of other women. However, the data indicated that women who participated in the vocational training programme were much younger and relatively more educated than the women in the comparison group. Thus it was not clear whether the lower fertility among the participants of the vocational training programme was due to their relatively lower age and higher level of education or due to the effect of the project. Another example of how selection bias can give rise to apparently contradictory results can be found in a study undertaken in Indonesia (reported in Hull and Hull, 1988). The study found that an increase in the use of family planning among a group of women who were trained in income-generating activities was not significantly higher than that of the women who did not take part in such activities. This apparent anomaly in the findings is not surprising since the required condition for women being accepted into the training programme was that they should be family planning acceptors. Perhaps in this case it was futile to assess the impact of the project on family planning use.

Although it is difficult to control completely for all potential sources of selection bias, such a bias can often be minimized. A development project may not always be able to cover the entire target population in a given time period. Thus the population not yet covered by the project, which may possess similar characteristics to that covered, can serve as a comparison group in assessing demographic impacts.

For demographic impact assessment, care should also be taken in *choosing the experimental area* itself as some areas may benefit from the project while others may be adversely affected. The purpose of an impact assessment study should therefore be to assess the overall net effect of the project. A study of the effect of an irrigation project in Thailand (Prasaratkul *et al.*, 1985) found that fertility in the irrigated villages was lower than that in the non-irrigated villages. But a closer look at the available data and the methodology adopted, however, revealed that the non-irrigated villages, which were selected as a control group, were adversely affected by the irrigation project. In such a situation, the experimental group should have included areas affected by the project both beneficially and adversely and the control group should have included areas not affected by the project in any way.

Another methodological issue that needs careful attention is that of *choosing and defining appropriate variables* for measuring demographic changes. In evaluating the impact of development projects, the usual fertility measures used in descriptive studies such as children ever born may not always be appropriate. Measures must therefore be developed carefully to capture changes during the reference period of the study. Thus in evaluating the impact of a project that has been implemented for five years only, the dependent variable or variables must be measured in such a temporal context, allowing for some time-lags where such variables can be expected to change. Failure to do so can often confound the results.

For example, the study on demographic impacts of a small-scale industry promotion programme in the Philippines by Pernia and Pernia (1986) found that the programme did not have any effect on lowering fertility. A closer examination of the methodology used by the study, however, revealed that the lack of a significant impact was perhaps due to the fact that the dependent variable used was the number of children ever born (CEB), which included events in the past that could not possibly be influenced by the project that started only a few years ago.

If due care is also not taken in the selection and specification of the independent variable or variables, the analysis can provide misleading results. For example, in the study by Seligson (1979) on the impact of land reform on fertility, it was found that land ownership had pro-natalist impact in Costa Rica. However, a careful scrutiny of the methodology adopted revealed that the independent variable 'land ownership' used was a catch-all term which included not only landowners (both titled and untitled) but also squatters, renters and share-croppers. The study therefore did not distinguish between the effects of land ownership and of 'landholding' on fertility.

The importance of *time-frame* in impact analysis should also be given due recognition. The effects of colonization/settlement programmes, for example, on migration can be fairly quick, because migrants respond promptly to economic opportunities. In the case of fertility, however, improvements in economic opportunities often require considerable time to have effects on the proximate determinants of fertility, perhaps a decade or more rather than a few years as assumed by most impact assessment studies reviewed in Chapter 2. Such lags should be carefully considered in designing surveys to collect information to assess the fertility impacts of development projects. It would be desirable to have a significant number of years, say about ten, between a baseline survey and a later survey. In the case of a single-round retrospective survey, at least a partial life history going back to the time before the project was implemented needs to be constructed.

In general, the length of the time-frame built into a research design should depend on the time-lag between the initiation of a development project and the expected occurrence of the demographic effects. If too short a time-frame is selected, the demographic effects may not yet have manifested themselves. If the time-frame is too long, the experiment may be contaminated. Contamination may occur as a result of either spillover of development inputs into the control areas – if these are adjacent to the experimental areas – or other exogenous changes which occur over time. In such a case, the methodological approach should involve an experimental/control design and use of longitudinal data. But an ideal application of this approach requires that:

1 the project be implemented in a large number of randomly selected areas;
2 there be no other concurrent development interventions in the area;
3 information be collected before and after introduction of a project.

If requirement (1) is met, areas receiving the project may be compared with those not receiving it. The former areas can serve as a control group even if requirement (2) is not met, as long as the other interventions are randomly distributed and their effects on target variables such as fertility and migration are smaller than those of the project under investigation. But requirement (3) adds a time dimension that may lead to contamination of the experiment. Isolating demographic effects of the project under study from changes introduced by other development interventions then becomes very difficult.

There is also the problem of distinguishing between short-term and long-term effects. An evaluation conducted at one point in time does not necessarily indicate a long-term trend, and considerable caution is required in drawing conclusions.

While the existing theoretical and empirical literature may provide general hypotheses for determining impact, it is still useful to undertake initial *exploratory studies* regarding the potential effect of a development project to identify additional and more specific hypotheses, and to determine the context in which certain relationships are expected to occur. The results of these exploratory studies can provide specific guidance on collecting further data and on developing a conceptual framework. Above all, they can suggest whether or not it is worthwhile carrying out a detailed impact assessment. For example, in the study by Uyanga (1985) on the demographic effects of the Cross River plantation project in Nigeria, no significant difference was found in fertility behaviour among plantation and non-plantation populations. However, since the social factors in the areas were found to be such that the rural women associated their status with the number of children they had given birth to, it was not surprising that the Cross River plantation project did not have any significant impact on fertility. The attempt to assess the impact of the project on fertility was therefore futile. A pilot study on the attitude of the women towards ideal family size prior to the impact assessment survey would have been sufficient to indicate that no impact of the project on fertility could be found.

*Demographic impact assessment should be broad based,* covering both demographic and socio-economic effects. Project outputs cannot produce demographic impacts if they have not yet produced socio-economic effects. A narrowly defined impact evaluation may conclude 'absence of any impact'. But such an assessment cannot suggest whether the lack of impact is due more to incomplete or faulty implementation of the project itself or to inappropriate intervention. Since development projects have specific non-demographic objectives which are valid in their own right, demonstration of 'no demographic impact' is not likely to lead to decisions to abandon the project or to modify project components merely to achieve demographic goals. For demographic impact studies to be more useful, therefore, they should also address the question of whether or not the project is producing the expected effects for which it was designed, and if not, why not. In doing so, lack of demographic impact can more easily be traced either to lack of effects or to lack of inherent effectiveness of the project in significantly influencing the socio-economic and proximate determinants of demographic outcomes. The additional information not only will provide a stronger basis for proper interpretation of impact results, but could also help to identify project components that could be modified to intensify their impacts on demographic variables as well as to improve the overall performance of the project.

In the Rodmanee and Bunnag (1983) study on a rural job creation programme in Thailand, for example, the programme was found to have no impact in deterring rural out-migration. It was found that the programme had provided negligible financial benefits to those who had participated in it. It provided poor rural workers with work for little more than ten days in a year and those who participated earned, on average, only 69 baht a day. Thus the lack of demographic impact observed in this study was perhaps due to the relatively negligible size of the development intervention.

Another example of a lack of inherent effectiveness of a programme due to the small size of the development intervention is provided by the study by Muhuri and Rahman (1982). In this study, as discussed earlier, 59 per cent of women who were members of co-operatives earned no income and only 21 per cent earned, on average, about 100 taka per month. It was not therefore surprising that the programme did not have any effect on fertility behaviour of this group of women.

The broader impact assessment will, however, require *a wider range of information* both at household and community level. And since a development project often comes with additional infrastructure, information on these other types of infrastructure (including dates of installation and extent of usage) is also crucial.

In most developing countries, the registration system of vital events is deficient, and information about demographic behaviour collected retrospectively or prospectively has its own problems and deficiencies. It is therefore often difficult to assess the impacts of a development project using readily available data. In impact assessment, while small-scale and inexpensive pilot studies can collect useful data and lead to testable hypotheses, they cannot yield unambiguous results. Therefore, there is a need to allocate sufficient funds at the outset of a development project to collect baseline data in both experimental and control areas, to be followed some five to ten years later by a separate survey. A survey in only experimental areas is not sufficient. The alternative is a large-scale, single-round, *expost facto* survey with intensive, retrospective data collection in both experimental and control areas to inquire about conditions prior to the project implementation as well as those prevailing at the time of interview. While the latter approach has the advantage of not requiring a long funding commitment, it may (a) encounter memory recall problems of respondents, and (b) suffer distortions because of socio-economic changes that may have occurred in the project area resulting from factors other than the project.

But whatever the methodology and research design used, one has to be careful in generalizing the results of the impact assessment study. The extent to which the results can be generalized depends on the representativeness of the study units. If the impact study is undertaken in only one or a few areas, it may not be correct to generalize the findings for the entire country. But undertaking a large national impact assessment study can also become unmanageable and much of the area-level contexts can be lost if the entire national sample consists of small study units dispersed all over the country. A better approach perhaps is to replicate highly focused impact assessment studies in a number of areas using similar research design. The cumulative information from all these studies can then form the basis for making generalizations with high levels of confidence. In a limited sense, the studies on the impact of rural electrification in Misamis Oriental and Cagayan Valley discussed in the previous chapter illustrate the gains that can be achieved from such replications.

And, finally, since it is often a combination of development inputs, rather than a single input in isolation, which is likely to have significant demographic effects, future impact assessment studies should attempt to identify the development inputs and their mixes which are most likely to generate large and rapid socio-economic and demographic changes in the communities concerned.

Thus instead of the traditional approach of determining the impact of a specific project, a research strategy that may be more useful will select areas where socio-economic and demographic changes have been rapid, and areas where such changes appear to be lagging. The study might then examine what development interventions and their mixes are present in each area and determine to what extent each set of development interventions, and each component of these sets, explain the observed differential in socio-economic and demographic changes between the two areas. This type of analysis can provide a more solid basis for formulating and implementing appropriate integrated area development programmes which would generate maximum socio-economic and demographic impacts.

Chapter 4

# Conclusions and policy implications

In spite of their methodological shortcomings, the case studies reviewed above do suggest certain guidelines for assessing demographic impact and for designing future development interventions so as to enhance the impact of such interventions on demographic behaviour.

First, it is crucial to assess the social environment under which a development intervention is planned in order to identify the factors that might neutralize or reduce the impact of the intervention on demographic behaviour. The case study on plantation settlement in Nigeria discussed above has shown that the fertility behaviour of the women in the plantation areas was almost identical to the high-fertility behaviour of the women in the non-plantation areas, although both these areas were provided with family planning services. A closer look at the evidence provided revealed, however, that in the study areas women's status was positively related to the number of children they had. Thus, even when the women made use of the family planning services, it was done to control the time of pregnancy rather than to reduce the family size. In this region, therefore, there is a need to change social attitudes, perhaps through investment in education. In fact the study has shown that fertility among educated women, both in the plantation and non-plantation areas, was lower than that of the women who had no education.

The analysis of the reforestation programme in Thailand suggests that female labour force participation was negatively related to recent fertility, which implied that the programme had helped to reduce fertility through employment of women in reforestation activities. On the other hand, the available evidence suggests that female labour force participation in home industry was not incompatible with high fertility. These results therefore suggest that policy-makers should pay greater attention to the *type of female employment* that needs to be generated to achieve greater anti-natalistic impacts. One alternative may perhaps be to promote non-agricultural employment opportunities in rural areas. But such an approach may not be cost-effective. Social cost-benefit analysis is therefore crucial for assessing these alternative policy approaches.

Since the socio-economic and demographic impacts of a specific project depend on the extent of productive use being made of the project inputs, one major task for the policy-makers should be to identify ways in which a large section of the population in the project area can be encouraged to make productive use of the services of the project. For example, the study on rural electrification in Thailand had found that only 0.2 per cent of households in the electrified village were using electricity for productive purposes (e.g. home industry, agricultural activities, animal raising, etc.). The study also suggested that the average villager could not afford the luxury of the various amenities which electricity can provide. It was not surprising, therefore, that electrification had no impact on fertility. Similar conclusions emerge from the two case studies in the Philippines, one in the Misamis Oriental region and the other in the Cagayan Valley. While in the Misamis Oriental region rural electrification was found to have a negative impact on fertility, no such relationship existed between fertility and rural electrification in the Cagayan Valley. Such differences in the impact of rural electrification are not difficult to explain. In the Misamis Oriental, rural electrification was a part of an overall development programme which enhanced the income and employment opportunities of the population, thus enabling them to use electricity more widely and productively. Also in this region, households as well as schools, health units and so on, were provided with financial help towards meeting the cost of installation and use of electricity. In the Cagayan Valley, on the other hand, the cost of electricity was found to be much higher than that in Misamis Oriental, as a result of which there was no significant growth in industry, employment and income in the region. The analysis therefore in general suggests that a rural electrification project with limited supply and high cost of electricity is not likely to achieve a wide coverage of the rural population nor high rates of productive use of electricity. As a result the project will have a limited impact on income, employment, health and sanitation, which have a bearing on fertility, mortality and migration outcomes. The task of policy-makers in such situations should therefore be to enhance the ability of the population of the project areas to make proper use of the project's input.

As development projects reduce the demand for children through changes in income and participation of women in the labour force, there may be an increasing demand for family planning services in project areas. Population policies should therefore be co-ordinated to meet the growing demand for such services in the wake of development interventions. The case study on rural electrification in Bangladesh noted that although rural electrification may have given rise to attitudinal changes among rural women in favour of a smaller ideal family size, the inadequate provision of family planning services was perhaps the main reason why the desire for a smaller family failed to be translated into a significant reduction in actual fertility.

On the other hand, there is also some evidence that family planning programmes implemented in isolation are unlikely to achieve more than moderate success in increasing contraceptive use and reducing fertility in situations where the economic value of children is high because of limited opportunities of parents for income generation and old age support, and in situations where infant and child mortality are high because of limited access to health and sanitation services. Thus, until improvements in income and living conditions reduce the need to have large families, there will be little demand for contraception.

An integrated population and development approach is thus necessary to enhance the demographic impacts of a particular intervention. In the case study on electrification in Java and Bali in Indonesia, it was observed that while there was a higher rate of acceptance of family planning services in rural areas (because of the community pressure and initial concentration of these services in such areas), the distribution of electricity tended to be biased towards urban and suburban areas, and heavily favoured better-off districts. In such situations, neither the benefits of electrification – as far as its impact on fertility behaviour is concerned – nor that of the family planning programmes can be fully realized.

The size of a development intervention ought to be carefully judged, because if an intervention is too small in size it may completely fail to generate the desired result. For example, one aim of a rural job creation programme in Thailand was to reduce rural–urban migration. But the available evidence suggests that the programme had completely failed to deter rural out-migration largely because it generated very little work (about seven days a year) and income (about 70 baht per day) for the participants. In a study of women's co-operatives in Bangladesh it was also noticed that the programme had generated very little income (up to 100 taka per month), and that too for fewer than half of the co-operative members. It is not surprising, therefore, that the fertility behaviour of the co-operative members was not affected by this particular intervention.

With regard to women's co-operatives, the evidence also suggests that economic activities associated with them are usually carried out near the home or do not conflict with women's child-bearing and child-rearing roles. Women's participation in such activities is therefore unlikely to motivate them to have fewer children. Indeed, one could plausibly argue the reverse. The programme therefore needs to be evaluated using a cost-benefit approach, and other options considered that are more cost-effective.

The finding that family planning users in Indonesia who were trained in income-generating schemes show consistently higher proportions still continuing to practise contraception and switching to permanent methods suggests that incentives are perhaps needed to sustain the acceptance of family planning, particularly where acceptance levels are already very high.

The analysis of the land reform programme in Thailand shows that the impact of the programme in reducing migration and fertility has been greater when other development inputs such as technical and financial assistance for small farmers were made available in land reform areas. This suggests that perhaps an integrated development approach is more effective in improving living standards and fulfilling national demographic objectives than a single-purpose development intervention.

It should, however, be mentioned that while a land redistribution programme is likely to depress fertility, it may result in increased fertility, at least in the short run. For example, Seligson's (1979) study in Costa Rica noted higher fertility among landed households than among landless households. As discussed earlier, this may happen for two reasons. First, households which experience increases in the size of their holdings under a redistribution programme can more profitably utilize child labour on their increased holdings. Second, the higher incomes associated with larger holdings can be expected to increase natural fertility and the survival prospects of children. But such a response of fertility to a land redistribution programme is unlikely to hold in the long run, particularly if rising incomes alter educational expectations for children. Moreover, redistribution programmes rarely take place in the absence of some type of tenurial reform. If a land redistribution programme is accompanied by changes in land ownership, the pro-natalist impact of the redistribution programme may be partially or totally offset. The increased ownership is likely to depress fertility as land substitutes for children as the primary source of security in old age. More work is, however, needed to distinguish clearly between the impact on fertility of the size of holdings to which the household has access for cultivation and the amount of land that is owned.

Nevertheless, the potential impact of land redistribution on rural development and fertility is sharply limited by the supply of available land for redistribution. In some areas, such as large parts of Asia, the physical availability of land provides the constraint in that where population pressure on land is high, a forced redistribution would result in a large number of uneconomic holdings. In such cases, alternative strategies such as guaranteed employment schemes may be more appropriate to influence the economic conditions and fertility of landless labourers. In other areas of the world, land is not in short supply but political factors prevent its redistribution. Thus, even in many countries where laws on ceilings for landholdings have been enacted, progress in redistributing land has been slow.

One of the objectives of irrigation projects in Thailand was to reduce migration to the urban centres. But the evidence suggests that two of the three irrigated villages had actually lost people through out-migration. Normally, one would expect the higher incomes due to increased productivity of land, and the increased employment opportunities associated with an irrigation project, to slow down migration out of the project areas. This, however, depends on whether it is small farmers or large farmers who benefit from the project. The analysis shows that by increasing productivity of land irrigation had also given rise to an increase in land prices in two of the irrigated villages, thereby making the cost of farming probably too high for the poor farmers. Moreover, irrigation projects are usually accompanied by other development inputs such as new roads which open up contacts with urban centres and thus facilitate migration. Thus, policy-makers should give due attention to the possible income distribution effects of a development intervention and its potential effect on demographic factors.

Policy-makers should assess both the positive and negative effects of a particular development intervention and adopt appropriate measures to mitigate negative consequences. For example, while an irrigation project such as building a dam may eventually bring benefits in terms of increased agricultural production in some areas, other areas may be adversely affected by losing their source of natural irrigation. This may cause out-migration from the areas adversely affected by the project due to loss of income and employment opportunities. Hence the policy-makers should make provisions, either through resettlement programmes or creation of alternative employment opportunities, so that the population of the areas adversely affected by the irrigation project have alternative sources of income to fall back on.

The evaluation of land settlement projects in several countries such as Brazil, Indonesia and Somalia shows that their impact on fertility can be pro-natalist, particularly where allocation of land leads to increased demand for family labour. This may suggest that if it is desired to reduce the fertility of women living in project areas, alternative policies which reduce demand for family labour and an active family planning programme would need greater attention. One such alternative policy could be the use of a more capital-intensive technology in settlement areas. But this may defeat the other objective of promoting employment. Thus, as argued earlier, it may not always be possible or even desirable to pursue anti-natalist policies, particularly if there is a conflict between the objectives of reducing population growth and achieving other welfare goals such as an increase in agricultural production and employment promotion. Nevertheless, such conflicts need to be taken into account in the cost-benefit analysis used to judge the relative significance of alternative policies and programmes or in designing projects to minimize adverse consequences on population growth.

Overall, the case studies reviewed suggest in general that the socio-economic and demographic effects of a specific development project are often strengthened in the presence of other development interventions and availability of family planning services. Therefore, the more integrated a development intervention is, the greater are its chances of success. The experiences of the Taepa Self-Help FELDA Scheme in Malaysia and the Bicol River Development Project in the Philippines show that, for socio-economic effects in general and demographic effects in particular, the integration of development projects with family planning projects is perhaps the most effective strategy. Furthermore, in view of the findings that development projects take some time to influence demographic behaviour, it would be useful for policy-makers to assess the immediate impact of the projects on intermediate variables which are often more easily monitored than levels of fertility and mortality. It can then be assumed that changes in these variables would have a predictable – even if unmeasurable in the short term – impact on demographic outcomes. Such a strategy has the advantage of avoiding some of the pitfalls faced by impact assessment studies in many countries.

# Notes

## 1 Introduction

1 The decline in mortality in developing countries was largely due to the use of modern public health measures and imported medicines such as antibiotics.

## 2 Demographic impact of specific development projects

1 The variable 'banana zone' was introduced to take into account the effect of the greater than average number of bachelor males who work in settlements located in areas with large banana plantations.
2 Preferred family size relates to what a person feels he/she should have whereas ideal family size refers to what is generally considered as appropriate.
3 *Agregados* are people who rent, share-crop or work for wages for the landowner and live on his plot.
4 MASICAP/SBAC-assisted enterprises received pre-loan consultancy service, loans at below the market rate of interest and post-loan technical assistance.

# Appendix: Salient features of studies reviewed in Chapter 2

*Table I* Studies on demographic impacts of electrification

| *Place/date of study* | *Study design and sample size* | *Measure of electrification* | *Years electrified* | *Perceived or actual effects of electrification* |
|---|---|---|---|---|
| THAILAND | | | | |
| Piampiti *et al.* (1982) | ● Study design: 'with and without'<br>● 5,000 households from 200 villages with electricity and 100 villages without electricity | ● Proportion of households electrified<br>● Years since village electrified | ● 1–25 years<br>● About two-thirds villages electrified less than 5 years | ● Direct effects not clear<br>● Indirect effects through female labour force participation, husband's income, desired family size, child mortality and contraceptive practice |

| *Other development inputs (interventions)* | *Demographic variables used in impact assessment* | *Method of analysis* | *Results* | *Remarks/major shortcomings* |
|---|---|---|---|---|
| Differences existing at the time of survey taken into account | Village level: ● crude birth rate ● migration rates Household level: ● average number of children ever born ● number of children under age 3 ● current use of contraceptives ● number of children died | ● Multiple regression ● Multiple classification analysis ● Path analysis | Village level: ● electrification not significantly related to birth rate ● reduces out-migration Household level: ● no difference in fertility or infant mortality ● in electrified households 69 per cent use contraceptives compared to 63 per cent in non-electrified households ● use of electricity negatively related to labour force participation of women (result not in expected direction) | ● Separate regressions run for electrified and non-electrified villages. Electrification not included as an independent variable. Difficult therefore to interpret the effect of electrification ● Since two-thirds of villages electrified less than 5 years, too early to find impact on fertility ● Since electrified villages are better served by business, health and educational facilities, 'before and after' methodology would have been more appropriate. But since baseline information not available 'then and now' strategy should have been used to complement 'with and without' design |

*continued overleaf*

*Table I* – continued

| *Place/date of study* | *Study design and sample size* | *Measure of electrification* | *Years electrified* | *Perceived or actual effects of electrification* |
|---|---|---|---|---|
| THAILAND<br>Pekanan (1982) | ● Study design: 'with and without'<br>● 609 households: 307 households in electrified area, 302 households in non-electrified area | Presence or absence in village | 4 years | ● Not used for irrigation (since rainfed agriculture)<br>● Little economic use by households<br>● Increase in TV viewing |
| INDONESIA<br>1976<br>(Freedman *et al.*, 1981) | ● Study design: 'with and without'<br>● 4,243 villages: 3,491 non-electrified, 752 electrified | Presence or absence in village | | |

| *Other development inputs (interventions)* | *Demographic variables used in impact assessment* | *Method of analysis* | *Results* | *Remarks/major shortcomings* |
|---|---|---|---|---|
| Villages picked for similarity of other infrastructure | ● Live births in 4 years before electricity and in 4 years after access to electricity<br>● Use of contraceptives | ● Cross-tabulations<br>● Multiple classification analysis<br>● Analysis of variance | ● No significant difference in percentage use of contraceptives<br>● Recent fertility significantly lower in electrified villages | ● Conclusion that recent fertility lower in electrified villages is intriguing since no change in social conditions and economic status of households observed<br>● Only 4–5 per cent of households use electricity for economic purposes<br>● Perhaps the study has picked up spurious statistical relationship with fertility since 4 years is not long enough for electrification to influence fertility behaviour |
| Other infrastructure variables (health, education, etc.) included in the analysis | Use of contraceptives | Multiple regression | ● Contraceptive use higher in non-electrified than in electrified villages (39 per cent as against 29 per cent)<br>● Lack of relationship between electrification and fertility | ● Electrification had limited spread and bias towards urban areas in 1970s (only 6 per cent of all households in 1971 and 14 per cent in 1980 had electricity)<br>● Contraceptive use was biased towards rural areas.<br>These factors have confounded the results of the study |

*continued overleaf*

*Table I* – continued

| *Place/date of study* | *Study design and sample size* | *Measure of electrification* | *Years electrified* | *Perceived or actual effects of electrification* |
|---|---|---|---|---|
| BANGLADESH 1983 (Robinson *et al.*, 1984) | ● Study design: 'with and without' ● 600 households in 24 villages with access to electricity: 400 electrified households, 200 non-electrified households | Presence or absence in household | 3 years (electrified in 1980–1) | |
| PHILIPPINES Misamis Oriental, 1980 (Herrin, 1988) | ● Study design: 'with and without' ● 1,402 households in 36 villages (18 each from western and eastern part of the Province) | Number of years community/ households electrified | 0–9 years (electrification introduced in western part in 1971 and in eastern part in 1978) | ● Increase in irrigation ● Growth of large-scale enterprises ● Wider availability of potable water |

| *Other development inputs (interventions)* | *Demographic variables used in impact assessment* | *Method of analysis* | *Results* | *Remarks/major shortcomings* |
|---|---|---|---|---|
| Not relevant for household comparisons within same villages | ● Completed family size ● Ideal family size ● Current use of contraceptives | ● Tabulations ● Path analysis | ● Electrified households had higher completed family size than non-electrified (5.9 compared to 5.4) ● Desired family size was lower among electrified than non-electrified households (2.9 compared to 3.2) ● No difference in contraceptive use ● Results of path analysis show electrification increasing use of family planning by enhancing husband–wife communication, access to information, outside home contacts and ability to earn income, etc. | ● 3 years is too short a period for evaluation of the effects of electrification ● Children ever born not an appropriate measure for assessing impact in the present context ● Study of fertility behaviour after households received electricity would have been more appropriate ● Comparison group also not appropriate since electrified households are richer and more educated |
| Controlled for other development inputs and community characteristics (measured in community survey) | ● Fertility in 5 or 2 years prior to survey ● Current use of contraceptives ● Child deaths of women under 35 years | Multivariate analysis to control for bias in selection of villages and households for electrification | Years community electrified related to lower fertility, greater use of family planning and lower child mortality | |

*continued overleaf*

*Table I* – continued

| *Place/date of study* | *Study design and sample size* | *Measure of electrification* | *Years electrified* | *Perceived or actual effects of electrification* |
|---|---|---|---|---|
| Cagayan Valley, 1980 (Herrin, 1988) | ● Study design: 'with and without' ● 600 households in 12 electrified and 16 non-electrified villages ● Non-electrified villages chosen using same criteria as used for original selection of villages for electrification | Presence or absence of electricity in community/household | 0–5 years | Little use of electricity for operating machines or in health services and schools |
| COLOMBIA 1980 (ter-Wengel, 1985) Data for period 1970–9 | ● Study design: 'with and without' ● Total of 200 households in two villages (one electrified and the other partially electrified but with extended school programme for two years) | Presence or absence in village | ● 5 years in village ● Not collected for households | |

| *Other development inputs (interventions)* | *Demographic variables used in impact assessment* | *Method of analysis* | *Results* | *Remarks/major shortcomings* |
|---|---|---|---|---|
| Controlled for other factors measured in community survey (little difference between electrified and non-electrified areas) | • Fertility in 4 years prior to the survey<br>• Current use of contraceptives | Multivariate analysis controlling for area and household characteristics | No significant effect on demographic variables | No significant impact observed because use of electricity not widespread partly due to high cost (only 27 per cent of households had electricity) |
| | Migration decisions of persons in age-group 15–29 | Multivariate regression (time series data) | Electrification reduces out-migration | • Control group not well specified<br>• Not clear whether electrification or extension of school programme reduced migration<br>• Choice of time-period not appropriate. Two years is too soon to study the effect of education on out-migration |

*Table II* Studies on demographic impacts of irrigation and integrated rural development programmes

| *Place/date of study* | *Study design and sample size* | *Measure of irrigation/ project participation* | *Years irrigated/ project implemented* | *Perceived or actual impacts of irrigation* |
|---|---|---|---|---|
| INDIA | | | | |
| Girna Irrigation Project in Maharashtra State (Mukerji *et al.*, 1986) | ● Study design: 'with and without' ● 100 villages: 50 experimental, 50 control group ● Based on subdistrict level data for 1971 and 1981 censuses | Presence or absence of irrigation at subdistrict level | | |
| THAILAND | | | | |
| Lam Pra Plerng, 1980 (Prasaratkul *et al.*, 1985) | ● Study design: 'with and without' ● 5 villages: 3 irrigated, 2 non-irrigated ● Household survey and anthropological approach used for collection of information on population change and development | Presence or absence in village | 8–10 years | Increased agricultural productivity, multiple cropping, higher land prices, solved annual flooding problem, provided clean water, improved toilet facilities |

| *Other development inputs (interventions)* | *Demographic variables used in impact assessment* | *Method of analysis* | *Results* | *Remarks/major shortcomings* |
|---|---|---|---|---|
| | Crude birth rate | Tabulations | ● Fertility remained high during 10-year period in both experimental and control areas ● Subdistrict which benefited most had highest birth rate ● From birth registration data, trend in annual birth rates downward, decline larger in experimental areas ● Concerning migration, experimental areas gained population through in-migration | Based on data not specifically collected for the present study ● to assess demographic effects, study depended on indirect estimates. In case of fertility, contradictory results obtained ● with regard to migration, poor civil registration data did not permit in-depth analysis |
| Expansion of roads, electrification and agricultural extension services | ● Migration ● Deaths (1975–9) ● Children ever born | Tabulations of information collected in: ● surveys ● participant observations ● interviews with key informants | ● Lower rates of out-migration in irrigated than in non-irrigated areas ● Irrigated villages lost population from lower in-migration due to higher land prices ● Lower fertility in irrigated than in non-irrigated villages (3.4 compared to 4.3) except for one village | ● Choice of experimental and control villages close to each other inappropriate since building of dam has aggravated conditions for agriculture in one of the control villages. Rising land prices have also affected control villages. These adverse effects should have been taken into account in assessing overall impact of irrigation ● Effect of patterns and size of land ownership on migration and fertility should also have been taken into account |

*continued overleaf*

*Table II* – continued

| *Place/date of study* | *Study design and sample size* | *Measure of irrigation/ project participation* | *Years irrigated/ project implemented* | *Perceived or actual impacts of irrigation* |
|---|---|---|---|---|
| Nam Pong Project (north-East) artificial lake (Supapong-opichate and Sangsritat-ankul, 1982) | ● 374 households in 12 villages ● 127 use ground water ● 217 never used and 30 past users | Number of years used ground water | | Permits multiple cropping and leads to more use of female labour |
| North-eastern region (Prasith-rathsint *et al.*, 1981) | ● Study design: 'with and without' ● 225 villages: 45 villages from each type of irrigation system (large, medium, small); 90 villages with no irrigation ● 4,500 households/ eligible couples (20 from each of 225 villages) | ● 4 types of irrigation system (large, medium, small and none) ● Size of irrigation defined in terms of total cost of project and area served by it | Not considered | Increased productivity and employment in dry season |

| *Other development inputs (interventions)* | *Demographic variables used in impact assessment* | *Method of analysis* | *Results* | *Remarks/major shortcomings* |
|---|---|---|---|---|
| | ● Migration<br>● Recent fertility | Tabulation:<br>● multiple classification analysis<br>● controlled for other variables: users tend to be better educated and have higher incomes than non-users | ● Out-migration from project area higher than in-migration (0.4 persons compared to 0.06 persons per household)<br>● Fertility and ideal family size lower among those benefiting from project, but when controlled for other independent variables those not benefiting have smaller desired family size and also do not want more children | Study did not have any control or comparison group |
| ● Villages with no irrigation less likely to have electricity or family planning unit<br>● At village level controlled for use of electricity, school and health station | ● Number of live births<br>● Live births in previous 5 years<br>● Crude birth rate<br>● Ideal family size<br>● Use of contraceptives | ● Multiple regression<br>● Path analysis | Household level:<br>● No clear relationship between irrigation and fertility. In total sample, only moderate participation in irrigation negatively related to fertility. Within each stratum, high, medium and low participation in irrigation is positively related to fertility. Electricity use reduces fertility in villages without irrigation<br><br>Village level:<br>● Size of irrigation negatively related to fertility only when all villages considered together<br>● No clear relationship between irrigation and fertility at village level | It would have been more useful to categorize villages by the duration of their use of irrigation since it takes time for an irrigation system to influence socio-economic and demographic factors |

*continued overleaf*

*Table II* – continued

| *Place/date of study* | *Study design and sample size* | *Measure of irrigation/ project participation* | *Years irrigated/ project implemented* | *Perceived or actual impacts of irrigation* |
|---|---|---|---|---|
| MALAYSIA<br>Integrated population–development programme<br>● FELDA (Federal Land Development Authority Programme) (Fong, 1983) | ● 26 FELDA schemes were selected<br>● 1,641 settler families for household survey<br>● 262 scheme officials for 'staff survey' | Performance of FELDA schemes measured by family planning, health, education and economic indicators | Not reported | Settlement of rural landless, land development, improvement in income and employment opportunities, better social services |
| PHILIPPINES<br>Bicol River Basin Development Project, 1975 (Herrin, 1988) | ● 1,906 households in 100 rural and urban communities<br>● Data from the Bicol Multi-purpose Survey (BMS) conducted in 1978 | Presence or absence at the time of the survey | 3 years (1975) | ● Higher wage rates for husbands and wives<br>● Development of water resources, expanded drainage, flood control<br>● Irrigation and provision of safe drinking-water supply |

| *Other development inputs (interventions)* | *Demographic variables used in impact assessment* | *Method of analysis* | *Results* | *Remarks/major shortcomings* |
| --- | --- | --- | --- | --- |
| Education, family planning services, agricultural extension | • Proportion of women practising family planning • Proportion of women using post-natal care • Proportion of 5–6 year-olds attending kindergarten • Proportion of husbands and wives in income-generating activities | Tabulations included stratification of variables to assess performance: • length of residence • type of crop grown in community • age of housewife | • 55 per cent eligible women practise family planning compared with national rural average of 35.5 per cent • 79 per cent use post-natal care compared with national rural average of 30 per cent | • Study did not have any control group • Results compared only with national average |
| Roads, electrification, agricultural services, health, nutrition and family planning | • Number of births in 5 years prior to survey • Current use of contraceptives | Multiple regression | • Rural electrification and irrigation are related to lower fertility; irrigation and travel time to the town centre have a significant negative effect • Electrification, irrigation and travel time to town have no effect on current fertility preferences (additional children desired), but these factors are significantly related to current family planning use | • In the absence of 'control' or 'comparison' group, difficult to determine how changes in fertility and family planning behaviour are different from those in non-programme areas • Since project set up in 1975, it is doubtful whether it would have influenced fertility during the period 1973–8 |

*continued overleaf*

*Table II* – continued

| *Place/date of study* | *Study design and sample size* | *Measure of irrigation/ project participation* | *Years irrigated/ project implemented* | *Perceived or actual impacts of irrigation* |
|---|---|---|---|---|
| THAILAND | | | | |
| Taepa Self-Help Settlement, 1955 (Prasith-rathsint, 1987) | ● 650 couples<br>● 50 per cent living in settlement and 50 per cent in non-settlement villages<br>● Survey conducted in 1980<br>● Women covered under 25 years of age | Living in settlement or non-settlement villages | 0–25 years of settlement | Agricultural development, higher income and better social services |
| BANGLADESH | | | | |
| Comilla-Kotwali Thana (Sub-Division) (Khuda, 1985) field work Oct. 1979–Apr. 1981 | ● Household survey and anthropological approach to data collection<br>● 50 per cent households covered in one village | | 15–20 years since implementation of integrated rural development programme | Agricultural development, increased employment opportunities and social services, women's involvement in productive activities and family planning |

| *Other development inputs (interventions)* | *Demographic variables used in impact assessment* | *Method of analysis* | *Results* | *Remarks/major shortcomings* |
|---|---|---|---|---|
| Assistance in physical planning, provision of social services (education, health, etc.), integrated agricultural programmes, covering production, marketing loans, promotion of collective and co-operative organizations | Number of live births | Multiple classification analysis included information on place of birth length of stay in the settlement, etc. | ● Living in Taepa settlement has a negative effect on fertility ● Those born in settlement have lowest level of fertility, followed by those who migrated in when single and those who live outside | ● Study based on one-shot cross-sectional survey, comparing two dissimilar groups ● Control villages being too near the experimental area may have influenced the results |
| Agricultural extension, credit, co-operatives, development of rural infrastructure, education | ● Mean number of children ever born ● Age-specific marital fertility rates ● Crude birth rate ● Total fertility rate | Tabulations | Integrated rural development activities associated with lower fertility in Sreebollobpur village | ● Study again underlines the problems of not having baseline data ● In the absence of control group, fertility in experimental village compared with national surveys carried out in 1961, about the same time as integrated rural development project was initiated |

*Table III* Studies on demographic effects of land reforms and agricultural resettlement schemes

| *Place/date of study* | *Study design and sample size* | *Nature of land change* | *Years since implemented* | *Changes directly related to land change* |
|---|---|---|---|---|
| COSTA RICA (Seligson, 1979) Study of land status and fertility, 1973 | ● Comparison of landed with landless households ● Survey of 459 peasant households (274 landed, 185 landless) conducted in 1973 | ● Landed category included landowners, squatters, renters and share-croppers ● Landless included plantation workers, day labourers and migrant workers | | |
| Study of land reform and fertility, 1976 | ● 1976 survey ● 527 male heads beneficiaries of land reform ● 422 male heads non-beneficiaries of land reform from both rural and urban areas | Redistribution of land | | |
| Study of communal ownership of land and fertility, 1976 | 226 communal, 527 individual owners of land | Communal ownership of land vs. individual ownership of land | | |

| *Accompanying development inputs* | *Demographic variables used in impact assessment* | *Method of analysis* | *Results* | *Remarks/major shortcomings* |
|---|---|---|---|---|
| Family size differences controlled for age, income, education, etc. | Number of children ever born | Multiple classification analysis | Landed peasants have larger families than landless ones (5.95 as against 4.91) | ● Actual family size not an appropriate measure of fertility since it may be that child mortality is higher among landless ● 'Landowners' include renters and share-croppers. Therefore the study does not distinguish the effect of landownership from landholding status |
| | Ideal family size as a proxy for fertility behaviour | Analysis of variance | Ideal family size among landholding households was greater than among landless (4.3 as against 2.8) | ● The comparison group includes persons from urban areas. Since fertility in urban areas is generally lower, this may have biased the results ● Study should have examined actual fertility behaviour of beneficiaries of land reform during the post-participation period, rather than ideal family size |
| | Ideal family size | Analysis of variance ● co-variates adjusted for age, income, education, settlement area, etc. | ● Ideal family size among communal landowners lower than among individual owners ● The effect of redistribution of land from communal to individual ownership is pro-natalist | |

*continued overleaf*

*Table III* – continued

| *Place/date of study* | *Study design and sample size* | *Nature of land change* | *Years since implemented* | *Changes directly related to land change* |
|---|---|---|---|---|
| NORTH-EASTERN THAILAND<br>Agricultural land reform project (Plainoi *et al.*, 1982) | ●667 families in reform villages, adjacent villages and distant villages<br>●Proportion of families owning land: reform villages 90 per cent; adjacent villages 72 per cent; distant villages 68 per cent | Allocated land to landless farmers | Up to 6 years | Increase in labour utilization and income |
| INDONESIA<br>Resettlement of Javanese transmigrants in South Sumatra (Oey, 1981) | ●Study design: comparison of fertility behaviour of transmigrants with rural stayers<br>●1,000 households from three areas (two resettlement areas in South Sumatra and one origin area in rural Java) | Provision of land to landless households | | Availability of land, increase in income and labour utilization |

| Accompanying development inputs | Demographic variables used in impact assessment | Method of analysis | Results | Remarks/major shortcomings |
|---|---|---|---|---|
| Electricity and piped water which hardly existed in other villages | ● Children ever born<br>● Children born in past 3 years<br>● Preferred, desired and ideal family size | ● Tabulations<br>● Multiple classification analysis | ● Children ever born highest in reform villages (3.76), followed by adjacent (3.42) and distant villages (3.39)<br>● Children born in past three years highest in distant villages<br>● Results on desired and ideal family size are mixed; preferred family size highest and ideal family size lowest in reform villages | Differences in fertility and family size preferences are small among three types of village. Perhaps the study was carried out too soon after the land reform (less than 6 years) |
| Agricultural extension services, roads, schools and hospitals | ● Children ever born<br>● Complete marital histories obtained from women aged 25–44 | ● Tabulations<br>● Analysis of variance<br>● Fertility differences controlled for age, education, age at first marriage, etc. | ● Fertility among transmigrants was higher than those who stayed in origin areas<br>● Part of the higher fertility among trans-migrants was due to selection factors<br>● Increased importance of subsistence farming and allocation of land resulted in increased demand for family labour | At the time this study was conducted, family planning programmes had hardly reached the transmigration area. This may have exacerbated the effect of environmental factors on fertility |

*continued overleaf*

*Table III* – continued

| *Place/date of study* | *Study design and sample size* | *Nature of land change* | *Years since implemented* | *Changes directly related to land change* |
|---|---|---|---|---|
| NIGERIA<br>Cross River Plantation Project (Uyanga, 1985) | ● Study design: 'with and without'<br>● Survey carried out in 1979 covered 8 government plantations and 200 randomly selected villages from non-plantation areas<br>● Sample size 4,036 respondents, (2,116 from plantations and 1,920 from non-plantation villages) | | Plantations established during 1950s | Plantations established to encourage migration from densely populated south-western part to the eastern region |

| Accompanying development inputs | Demographic variables used in impact assessment | Method of analysis | Results | Remarks/major shortcomings |
|---|---|---|---|---|
| | ● Mean number of children ever born<br>● Number of living children<br>● Additional children desired<br>● Migration | Tabulations controlled for age and education | ● 88 per cent of respondents were in-migrants; 56 per cent came during 1950s, 26 per cent during 1960s and 17 per cent during 1970s<br>● 70 per cent of migrants were males aged 21–30 and came from densely populated areas<br>● No significant difference in fertility behaviour between plantation and non-plantation population<br>● Although family planning facilities readily available to those who wanted it, only 11.2 per cent of plantation and 8.3 per cent of non-plantation population practised family planning<br>● Plantation women practised family planning to control the timing of birth rather than to restrict family size | The desire to have larger families, in both plantation and non-plantation areas, is due to the prevailing social conditions according to which women's status is related to the number of children they have. Thus, in this case, it was not appropriate to look for evidence of a reduction in fertility as a result of this particular development intervention |

*continued overleaf*

*Table III* – continued

| *Place/date of study* | *Study design and sample size* | *Nature of land change* | *Years since implemented* | *Changes directly related to land change* |
|---|---|---|---|---|
| BRAZIL | | | | |
| Rondônia Colonization Project (Henriques, 1985) | Study based on household surveys in two project areas, Ouro Preto and Gy-Paraná (sample size not available) | Allocation of land to landless households | Ouro Preto project established in 1970, Gy-Paraná in 1972 | Provision of land to landless and agricultural development |

| *Accompanying development inputs* | *Demographic variables used in impact assessment* | *Method of analysis* | *Results* | *Remarks/major shortcomings* |
|---|---|---|---|---|
| Provision of social services, agricultural extension, education and health facilities | ● Age-specific and total fertility rates<br>● Migration | Tabulations | ● Attracting migrants to project areas was not a big problem. In 1977 Government even discouraged flow of in-migrants<br>● Even though fertility declined in two settlement areas during 1975–80, it was still high. Total fertility rate was around 7 which is the same as for rural Rondônia<br>● High fertility was due to large demand for family labour and high level of infant mortality | Study did not report fertility rates for settlers and *agregados* (labourers) separately. Since landholding patterns can influence fertility behaviour, separate information on fertility for the two groups would have been revealing |

*Table IV* Studies on demographic impacts of women's projects

| *Place/date of study* | *Study design and sample size* | *Measure of project participation* | *Years since project implemented* | *Perceived or actual effects of project* |
|---|---|---|---|---|
| INDONESIA<br>Rural Co-operative/ Income Generating Scheme (Bruce, 1985) | Study based on field visits to beneficiaries of the revolving loan funds in 1984 | | Scheme started in 1980 and evaluated in 1984 | Objective of the scheme was:<br>● to improve quality of life of rural women who practise family planning<br>● to increase acceptance, practice and continuation of family planning<br>● to enhance training and management capabilities of women leaders |
| Income-generating project in south Sulawasi (reported in Hull and Hull, 1988) | Evaluation team used qualitative techniques, including in-depth interviews, informal conversations and observations of a co-operative meeting. Data on family planning were collected from records kept by trained women leaders and National Family Planning Co-ordinating Board | | Project started in 1980 | Project concentrated on training a group of 120 women leaders to promote and sustain a series of income-generating activities in their communities |

| *Other development inputs (interventions)* | *Demographic variables used in impact assessment* | *Method of analysis* | *Results* | *Remarks/major shortcomings* |
|---|---|---|---|---|
| Training courses and scheme of revolving loan funds also provided to villagers with at least 35 per cent acceptance rate | ● Number of children ever born<br>● Family planning acceptance | Descriptive analysis | ● Despite insufficient data, study observed that some women had been able to increase their income. Increased income generally spent on family food and children's education, which in turn contributed to higher quality of life<br>● Improved status of women observed through their control of funds and their increased activity in the community<br>● Enhancement of training and management skills among women leaders was noticed | ● Evaluation hampered by data limitations, study relied primarily on impressions<br>● Data not available on family planning use to compare results between project and non-project areas<br>● Although the team concluded that the project had had an impact on women's roles and fertility, lack of relevant data made the evaluation task difficult |
| | Family planning acceptance and use | Descriptive evaluation | Evaluation revealed certain anomalies. Family planning use was observed to have risen eleven-fold in the region in which the project was located, but similar increases were also observed in non-project areas | Since women selected for training project were already successful family planning acceptors, it was not possible to test the direct impact of the training project on family planning behaviour |

*continued overleaf*

*Table IV* – continued

| *Place/date of study* | *Study design and sample size* | *Measure of project participation* | *Years since project implemented* | *Perceived or actual effects of project* |
|---|---|---|---|---|
| BANGLADESH<br>Women's Co-operative Programme/ Men's Co-operative Programme, 1980 (Muhuri and Rahman, 1982) | ● Study adopted quasi-experimental design which included 'before' and 'after' observations in experimental and comparison areas. The experimental area consisted of villages in *thanas* where integrated rural development programmes (IRDP) were implemented and comparison area included non-programme villages.<br>● A sample of 1,470 households was selected from 31 villages<br>● Altogether 1,278 currently married female and 963 male respondents were interviewed. Baseline data for 1976 available only for 14 villages | ● Presence or absence of IRDP<br>● Four types of village were selected:<br>1. with women's co-operatives only<br>2. with men's co-operatives only<br>3. with both men's and women's co-operatives<br>4. villages with no co-operatives | Programme implemented over the period July 1975 to April 1980 | |

| *Other development inputs (interventions)* | *Demographic variables used in impact assessment* | *Method of analysis* | *Results* | *Remarks/major shortcomings* |
|---|---|---|---|---|
| | ● Total fertility rate<br>● Children ever born | Tabulations | Village level:<br>● Maximum decline in fertility took place in non-programme villages<br>● Fertility had risen slightly in villages with women's co-operatives<br>Household level:<br>● Age-specific marital fertility among members of women's co-operatives and wives of members of men's co-operatives higher than among females not covered by co-operative programme<br>● Fertility among wives of members of men's co-operatives generally higher than other women in most age groups<br>● Women members of co-operatives do not have a large monthly income. Only 21 per cent have above 91 taka and their income constituted only 13 per cent of household income | ● Results need to be carefully interpreted since baseline data for 1976 not available by village type<br>● Both women's co-operative members and wives of members of men's co-operatives on average, belong to richer section of the community, which makes the findings a little more surprising<br>● Income generated through the programme is very small. Also four years is too short a period for full effects of such programmes to be felt upon income and fertility behaviour |

*continued overleaf*

*Table IV* – continued

| *Place/date of study* | *Study design and sample size* | *Measure of project participation* | *Years since project implemented* | *Perceived or actual effects of project* |
|---|---|---|---|---|
| Women's Vocational Training (WVT) Programme (Rahim and Mannan, 1982) | Respondents randomly selected from three groups: ● 521 ever-married (257 currently married) directly exposed to the programme ● 655 currently married, not directly exposed to the programme ● 312 currently married from non-programme villages ● Survey carried out in Sept.–Oct. 1979 | Exposure or non-exposure to the programme | Programme implemented in 1976 | Major objective of the programme was to improve women's socio-economic conditions by upgrading their skills and to prepare them for acting as agents for social change |
| Mothers' Club Programme (Siddiqui *et al.*, 1982) | ● 2 mothers' clubs from each of the four selected *thanas* ● 600 households (75 from each mothers' club) ● Selected households included 195 respondents who were direct participants (members of club) and 471 indirect participants (living in programme villages). Of 195 respondents, information related to only 113 currently married was used. Remaining 82 were unmarried ● Sample included 248 households from non-programme villages ● Survey carried out in 1980 | Exposure or non-exposure to the programme | 1–5 years; programme initiated in 1975 | To give vocational training and instruction in family planning and basic education |

| *Other development inputs (interventions)* | *Demographic variables used in impact assessment* | *Method of analysis* | *Results* | *Remarks/major shortcomings* |
|---|---|---|---|---|
| | ● Total fertility rates<br>● Age-specific fertility rates | Tabulations | ● Total fertility rate of women directly exposed to the programme (3.6) was the lowest in comparison to that of indirectly exposed (7.5) and women in non-programme villages | ● Lower fertility among directly exposed women may not be due to the programme since the programme had been in existence for only 3 years; directly exposed women are relatively younger than other women and relatively more educated |
| Access to contraceptive supplies and recreational facilities were also made available | ● Number of children ever born<br>● Age-specific fertility rates | Tabulations | ● Total fertility rate lowest among direct participants (4.7) compared to indirect participants (7.5) and non-participants (9.5)<br>● No statistically significant relationship found between programme exposure (number of births during past year)<br>● Mean number of children ever born: direct participants (4.4) and non-participants (4.0) | ● Study did not control for age, education and other characteristics of participants and non-participants<br>● Lower fertility among participants may be due to the differences in these factors rather than the effect of the programme. Direct participants are young and relatively better educated<br>● Exposure to the programme is also very short for any impact on fertility to be felt |

*Table V* Studies on demographic impacts of rural job creation programmes and promotion of small-scale industry

| *Place/date of study* | *Study design and sample size* | *Nature and type of programme* | *Years since implemented* | *Perceived or actual effects of the programme* |
|---|---|---|---|---|
| THAILAND<br>Rural Job Creation Programme (Rodmanee and Bunnag, 1983 | ● Study design: 'with and without'<br>● Sample size: 818 household heads from among members and non-members of the programme<br>● Survey carried out in Oct.–Nov. 1981 | | | ● Employment and income generation<br>● Reduction in seasonal out-migration |
| Rural Small-Scale Industry Promotion Programme (Chalamwong, 1983) | ● Survey carried out during Mar. 1980–Feb. 1981 covered 247 households in 22 villages in three provinces<br>● 261 currently married women in selected households were interviewed | | | ● Increased female labour force participation<br>● Higher income |

| *Other development inputs (interventions)* | *Demographic variables used in impact assessment* | *Method of analysis* | *Results* | *Remarks/major shortcomings* |
|---|---|---|---|---|
| | Seasonal out-migration | ● Tabulations<br>● Multiple classification analysis | ● Programme did not reduce seasonal out-migration from rural areas<br>● Participants had employment for very short period (7.4 days on average during dry season)<br>● Income earned from the programme was only a small proportion of total income | There is little point in searching for demographic impacts in situations where benefits generated by the programme are small |
| | Children ever born | ● Tabulations<br>● Regression analysis (two-stage least squares) | ● Non-linear relationship between fertility and female labour force participation in small-scale industry<br>● Number of children ever born peak between 1,500 and 2,000 work hours per year of the mother<br>● No significant relationship between fertility and number of years engaged in home industry<br>● Regression results indicate that women's participation has a negative effect on fertility except when participation is measured in terms of years engaged in home industry<br>● Child labour force participation has a small but positive effect on children ever born | The size of regression coefficients indicates that the negative effect of female labour force participation on fertility is small. When measured in terms of years engaged in home industry, female labour force participation has a positive effect on children ever born. The study discounts this result on technical grounds but female participation in home industry can give rise to increase in demand for children, particularly if they are used as labour in home industry |

*continued overleaf*

*Table V* – continued

| *Place/date of study* | *Study design and sample size* | *Nature and type of programme* | *Years since implemented* | *Perceived or actual effects of the programme* |
|---|---|---|---|---|
| PHILIPPINES<br>Small-Scale and Cottage Industry Programme, Bohol Province, 1980 (Pernia and Pernia, 1986) | Enterprise survey:<br>● 85 enterprises were covered<br>● 31 received assistance from small- and medium-scale industries programme<br>● 20 received assistance from other programmes<br>● 34 were unassisted<br><br>Household survey:<br>● Information obtained from 428 households. Of these, 299 were owner and hired-worker households assisted by some programmes and 129 were unassisted | Programme included assistance to small-scale industry (employing 5–99 workers) and cottage industries (employing 1–4 workers) | Assistance programme started in mid-1970s | ● Increased income and employment<br>● Increased health<br>● Greater labour force participation |
| INDIA<br>Rural Cottage Industry Programme in West Bengal (Bhattacharya and Hayes, 1983) | ● Study based on a survey of 362 families participating in cottage industry<br>● Fieldwork carried out in 1980<br>● Study design: participation in cottage industry | | | Increased income and female labour force participation |

| *Other development inputs (interventions)* | *Demographic variables used in impact assessment* | *Method of analysis* | *Results* | *Remarks/major shortcomings* |
|---|---|---|---|---|
| | Children ever born (CEB) | Regression analysis | ● Programmes tended to increase income, particularly of owner households ● Increased incomes had a negative effect on fertility after threshold level (4,000 pesos) | One should be cautious in interpreting the results: ● coefficients of threshold income variables are negligible and statistically insignificant ● dependent variable (CEB) includes events in the past that cannot be influenced by the current programme |
| | Children ever born | Tabulations | ● Women with children working in cottage industries have higher fertility than women whose children do not work in these industries ● But once children start working the former group has a lower fertility than the latter ● High fertility causes children's participation in cottage industries, but after participation it acts as a break on additional fertility | The ideal research design for this study would probably have been a panel study, following families through time looking at their work and fertility history in order to examine their fertility behaviour before and after they start participating in cottage industries |

# Bibliography

Barlow, Robin (1982) *Case Studies in the Demographic Impact of Asian Development Projects*, Center for Research on Economic Development, Ann Arbor, Michigan: The University of Michigan.

Bhattacharya, A.K. and Hayes, A.C. (1983) *Cottage Industry and Fertility in West Bengal, India*, paper presented at the 52nd Annual Meeting of the Population Association of America, Pittsburgh, April 1983 (mimeo).

Bruce, J. (1985) 'Income generating schemes: the Bangladeshi experience', *Populi* 12, 2: 38–49.

Cain, M. (1981) 'Risk and insurance: perspectives on fertility and agrarian change in India and Bangladesh', *Population and Development Review* 7, 435–74.

Chalamwong, Y. (1983) 'Development of cottage industries, women's labour force participation and fertility in rural Thailand', in S. Prasith-rathsint (ed.) *Population and Development Interactions in Thailand*, Bangkok: Prabpim.

Fong, C.O. (1983) *The FELDA Population and Development Program: Impact and Efficacy*, Faculty of Economics and Administration, University of Malaysia (mimeo).

Freedman, R., Khoo, S.E. and Supratilah, B. (1981) *Modern Contraceptive Use in Indonesia: A Challenge to Conventional Wisdom*, World Fertility Survey Scientific Reports, No. 2, London, World Fertility Survey.

Goldscheider, C. (1984) 'Migration and rural fertility in less developed countries', in W.A. Schutjer and C.S. Stokes (eds) *Rural Development and Human Fertility*, New York: Macmillan.

Harbison, S.F. and Robinson, W.C. (1985) 'Rural electrification and fertility change', in *Population Research and Policy Review* 4: 149–71.

Henriques, M.H.F.T. (1985) 'The demographic dynamics of a frontier area: the state of Rondônia, Brazil', in R.E. Bilsborrow and P.F. De Largy (eds) *Impact of Rural Development Projects on Demographic Behaviour*, Policy Development Studies No. 9, New York: UNFPA.

Henriques, M.H.F.T. (1988) 'Colonization experience in Brazil', in A.S. Oberai (ed.) *Land Settlement Policies and Population Redistribution in Developing Countries: Achievements, Problems and Prospects*, New York: Praeger.

Herrin, A.N. (1982) *The Cagayan Valley Rural Electrification Project: An Impact Assessment*, Diliman, Quezon City: Philippine Center for Economic Development.

Herrin, A.N. (1984a) 'Socio-economic-demographic impact: the case of rural electrification', in *Perspective for Population and Development Planning*, Manila: National Economic and Development Authority and Commission on Population.

Herrin, A.N. (1984b) 'Fertility and family planning behaviour in the Bicol River Basin', in *Transactions of the National Academy of Science and Technology*, vol. VI, Metro Manila: The Academy.

Herrin, A.N. (1988) *Demographic Impact of Development Projects: A Review of Selected Philippine Case Studies*, WEP 2–21/WP.162, Geneva: ILO.

Hull, V., Hull, J. and Singarimbum, M. (1977) 'Indonesia's family planing story: success and challenge', in *Population Bulletin* 32, 6: 1–48.

Hull, T.H. and Hull, V.J. (1988) *The Impact of Development Projects, Initiatives and Processes on Demographic Behaviour. Country Review: Indonesia*, WEP 2–21/WP.161, Geneva: ILO.

ILO (1984) *Population, Development, and Family Welfare: The ILO's Contribution*, Geneva: ILO.

Khuda, B. (1985) *Rural Development Programme in Bangladesh: Some Demographic Impact*, Liège, International Union for the Scientific Study of Population (mimeo).

Kocher, J.E. (1973) *Rural Development, Income Distribution and Fertility Decline*, New York: Population Council.

Kocher, J.E. (1984) 'Income distribution and fertility', in W.A. Schutjer and C.S. Stokes (eds) *Rural Development and Human Fertility*, New York: Macmillan.

Mabud, M.A. and Islam, A. (1987) *Impact of Community-based Development Projects on Fertility in Bangladesh*, study prepared for the ILO, Dhaka (mimeo).

McCawley, P. (1978) 'Rural electrification in Indonesia: is it time?', *Bulletin of Indonesian Economic Studies* 14, 2: 34–69.

Muhuri, P.K. and Rahman, M.B. (1982) *An Evaluation of the IRDP Pilot Project in Population Planning and Rural Women's Co-operatives, 1980*, External Evaluation Unit, Planning Commission, Government of the People's Republic of Bangladesh, Dhaka.

Mukerji, S., Kulkarni, S. and Audinarayan, N. (1986) *Demographic Consequences of a Development Project: A Study of Girna Irrigation Project, Maharastra State, India*, paper presented to a United Nations Workshop on Assessing the Demographic Consequences of Major Development Projects, New York, December (IESA/P/AC.21/9).

Oberai, A.S. (1987) *Migration, Urbanisation and Development*, WEP Background Papers for Training in Population, Human Resources and Development Planning, No. 5, Geneva: ILO.

Oberai, A.S. and Singh, H.K.M. (1983) *Causes and Consequences of Internal Migration: A Study in the Indian Punjab*, Delhi: Oxford University Press.

Oey, M. (1981) *The Impact of Migration on Fertility: A Case Study of Transmigrants in Lampung, Indonesia*, Thesis, Department of Demography, The Australian National University.

Pekanan, M. (1982) 'Impact of village electrification in northern Thailand on fertility', in S. Prasith-rathsint (ed.) *Fertility and Development Interactions in Thailand*, Rayong: Sahapromkarnpim.

Pernia, E.M. and Pernia, J.M. (1986) 'An economic and social impact analysis of small industry promotion: a Philippine experience', *World Development* 14, 5: 637–52.

Piampiti, S., Yongkittikul, T., Rodmanee, L. and Pasandhanatorn, V. (1982) *Demographic and Economic Impact of Rural Electrification in Northeastern Thailand*, Bangkok: Prabpim.

Plainoi, N. *et al.* (1982) 'The impact of agricultural land reform on fertility', in S. Prasith-rathsint (ed.) *Fertility and Development Interactions in Thailand*, Rayong: Sahapromkarnpim.

Population Council (1981) *Proceedings: Workshop on Methodological Approaches to Research on Fertility Impacts of Development*, Bangkok: Population Council.

Prasaratkul, P., Svetsreni, T., Dhanagom, C. and Pramularatna, A. (1985) 'Demographic impacts of rural development related to an irrigation project', in R.E. Bilsborrow and P.F. De Largy (eds) *Impact of Rural Development Projects on Demographic Behaviour*, Policy Development Studies No. 9, New York: UNFPA.

Prasith-rathsint, S. (1987) *A Review of Experiences with Development Projects and their Demographic Impacts in Thailand*, study prepared for the ILO, Bangkok (mimeo).

Prasith-rathsint, S., Srikhajorn, M., Arthorn-thurasook, T. and Pungpun, R. (1981) *The Impact of Agricultural Irrigation Projects on Fertility in Northeastern Thailand*, Bangkok: Vichit Hattakorn Press.

Rahim, M.A. and Mannan, M. (1982) *An Evaluation of the Women's Vocational Training Programme for Population Education and Control*, External Evaluation Unit, Planning Commission, Government of the People's Republic of Bangladesh, Dhaka.

Robinson, W.C., Begum, F.S. and Khatun, R. (1984) *The Impact of Electrification on Fertility in Rural Bangladesh: A Preliminary Report*, Policy Paper No. 6, Population and Development Planning Unit, Planning Commission, Government of Bangladesh, Dhaka.

Rodmanee, L. and Bunnag, S. (1983) 'Rural job creation program and migration in southern Thailand', in S. Prasith-rathsint (ed.) *Population and Development Interactions in Thailand*, Bangkok: Prabpim.

Seligson, M.A. (1979) 'Public policies in conflict: land reform and family planning in Costa Rica', *Comparative Politics* 12: 49–62.

Siddiqui, A.N.S.K., Ali, S., Rahman, M. and Chowdhury, A.H. (1982) *Evaluation of Use of Rural Mothers' Club in Population Activities*, External Evaluation Unit, Population Planning Section, Planning Commission, Government of the People's Republic of Bangladesh, Dhaka.

Stokes, C.S. and Schutjer, W.A. (1984) 'Access to land and fertility in developing countries', in W.A. Schutjer and C.S. Stokes (eds) *Rural Development and Human Fertility*, New York: Macmillan.

Supapongopichate, S. and Sangsriratankul, O. (1982) 'Fertility impact of man-made lake development: a case study of the Nam Pong Project, Northeastern Thailand', in S. Prasith-rathsint (ed.) *Fertility and Development Interactions in Thailand*, Rayong: Sahapromkarnpim.

ter-Wengel, J. (1985) 'The effects of electrification and the extension of education on the retention of population in rural areas of Colombia', in R.E. Bilsborrow and P.F. De Largy (eds) *Impact of Rural Development Projects on Demographic Behaviour*, Policy Development Studies No. 9, New York: UNFPA.

United Nations (1985) *World Population Trends, Population and Development Interrelations and Population Policies: 1983 Monitoring Report, Vols I and II*, New York: United Nations.

Uyanga, J. (1985) 'The demographic impact of the Cross River Plantations Project in Nigeria', in R.E. Bilsborrow and P.F. De Largy (eds) *Impact of Rural Development Projects on Demographic Behaviour*, Policy Development Studies No. 9, New York: UNFPA.

World Bank (1974) *Population Policies and Economic Development*, Baltimore and London: Johns Hopkins University Press.

World Bank (1984) *World Development Report 1984*, New York: Oxford University Press.

Zeidenstein, G. (1986) Foreword, in J. Stoeckel and A. Jain (eds) *Fertility in Asia: Assessing the Impact of Development Projects*, London: Frances Pinter.

# Index

For Product Safety Concerns and Information please contact our EU representative GPSR@taylorandfrancis.com Taylor & Francis Verlag GmbH, Kaufingerstraße 24, 80331 München, Germany

Printed and bound by CPI Group (UK) Ltd, Croydon, CR0 4YY
08/05/2025
01864370-0017